hands-on science

Level One

Jennifer Lawson

Joni Bowman

Randy Cielen

Carol Pattenden

Rita Platt

PEGUIS PUBLISHERS

Winnipeg • Manitoba • Canada

© 2000 Jennifer Lawson

Peguis Publishers acknowledges the financial support of the Government of Canada through the Book Publishing Industry Development Program (BPIDP) for our publishing activities.

Canadä

00 01 02 03 04 5 4 3 2 1

Canadian Cataloguing in Publication Data

Main entry under title:

Hands-on science : level one

    Ont. ed.
    Includes bibliographical references.
    ISBN 1-894110-44-7

1. Science – Study and teaching (Primary).
    I. Lawson, Jennifer E. (Jennifer Elizabeth),
    1959-

LB1532.L38 2000   373.3'5044 C00-920014-2

Series Editor:	Leigh Hambly
Assistant Editor:	Catherine Lennox
Book and Cover Design:	Suzanne Gallant
Illustrations:	Pamela Dixon,
	Jess Dixon

Program Reviewers

- Karen Boyd, Grade 3 teacher, Winnipeg, Manitoba
- Peggy Hill, mathematics consultant, Winnipeg, Manitoba
- Nancy Josephson, science and assessment consultant, Ashcroft, British Columbia
- Denise MacRae, Grade 2 teacher, Winnipeg, Manitoba
- Gail Ruta-Fontaine, Grade 2 teacher, Winnipeg, Manitoba
- Judy Swan, Grade 1 teacher, Winnipeg, Manitoba
- Barb Thomson, Grade 4 teacher, Winnipeg, Manitoba.

PEGUIS
PUBLISHERS

100-318 McDermot Avenue
Winnipeg, Manitoba, Canada R3A 0A2

Email: books@peguis.com
Tel: 204-987-3500 • Fax: 204-947-0080
Toll free: 1-800-667-9673

Contents

Introduction

Program Introduction

Hands-On Science develops students' scientific literacy through active inquiry, problem solving, and decision making. With each activity in the program, students are encouraged to explore, investigate, and ask questions as a means of heightening their own curiosity about the world around them. Students solve problems through firsthand experiences, and by observing and examining objects within their environment. In order for young students to develop scientific literacy, concrete experience is of utmost importance – in fact, it is essential.

The Foundations of Scientific Literacy

Hands-On Science focuses on the four foundation statements for scientific literacy, as established in the *Pan-Canadian Protocol.** These foundation statements are the bases for the learning outcomes identified in *Hands-On Science.*

Foundation 1: Science, Technology, Society, and the Environment (STSE)

Students will develop an understanding of the nature of science and technology, of the relationships between science and technology, and of the social and environmental contexts of science and technology.

Foundation 2: Skills

Students will develop the skills required for scientific and technological inquiry, for solving problems, for communicating scientific ideas and results, for working collaboratively, and for making informed decisions.

Foundation 3: Knowledge

Students will construct knowledge and understandings of concepts in life science, physical science, and earth and space science, and apply these understandings to interpret, integrate, and extend their knowledge.

Foundation 4: Attitudes

Students will be encouraged to develop attitudes that support responsible acquisition and application of scientific and technological knowledge to the mutual benefit of self, society, and the environment.

*Common Framework of Science Learning Outcomes K-12: Pan-Canadian Protocol for Collaboration on School Curriculum (1997).

Hands-On Science Expectations

UNIT 1: CHARACTERISTICS AND NEEDS OF LIVING THINGS

☐ Identify major parts of the human body and describe their functions.

☐ Identify the location and function of each sense organ.

☐ Classify characteristics of animals and plants by using the senses.

☐ Describe the different ways in which animals move to meet their needs.

☐ Identify and describe common characteristics of humans and other animals that they have observed, and identify variations in these characteristics.

☐ Describe some basic changes in humans as they grow, and compare changes in humans with changes in other living things.

☐ Describe patterns that they have observed in living things.

☐ Select and use appropriate tools to increase their capacity to observe.

☐ Ask questions about and identify some needs of living things, and explore possible answers to these questions and ways of meeting these needs.

☐ Plan investigations to answer some of these questions or find ways of meeting these needs.

☐ Use appropriate vocabulary in describing their investigations, explorations, and observations.

☐ Record relevant observations, findings, and measurements, using written language, drawings, charts, and concrete materials.

☐ Communicate the procedures and results of investigations for specific purposes, using demonstrations, drawings, and oral and written descriptions.

☐ Compare the basic needs of humans with the needs of other living things.

☐ Compare ways in which humans and other animals use their senses to meet their needs.

☐ Describe ways in which people adapt to the loss or limitation of sensory or physical ability.

☐ Identify a familiar animal or plant from seeing only a part of it.

☐ Describe ways in which the senses can both protect and mislead.

☐ Describe a balanced diet using the four basic food groups outlined in Canada's Food Guide to Healthy Eating, and demonstrate awareness of the natural sources of items in the food groups.

☐ Identify ways in which individuals can maintain a healthy environment for themselves and for other living things.

UNIT 2: CHARACTERISTICS OF OBJECTS AND PROPERTIES OF MATERIALS

☐ Identify each of the senses and demonstrate understanding of how they help us recognize and use a variety of materials.

☐ Describe various materials using information gathered by using their senses.

☐ Identify properties of materials that are important to the purpose and function of the objects that are made from them.

☐ Describe, using their observations, ways in which materials can be changed to alter their appearance, smell, and texture.

☐ Sort objects and describe the different materials from which those objects are made.

☐ Demonstrate ways in which various materials can be manipulated to produce different sounds, and describe their findings.

☐ Design a usable product that is aesthetically pleasing, and construct it by combining and modifying materials that they have selected themselves.

☐ Ask questions about and identify needs and problems related to objects and materials, and explore possible answers and solutions.

☐ Plan investigations to answer some of these questions or solve some of these problems.

☐ Use appropriate vocabulary in describing their investigations, explorations, and observations.

☐ Record relevant observations, findings, and measurements, using written language, drawings, charts, and concrete materials.

▶

- Communicate the procedures and results of investigations for specific purposes, using demonstrations, drawings, and oral and written descriptions.
- Describe how properties of materials help us learn about natural and human-made objects.
- Identify materials that can be used to join and fasten other materials.
- Demonstrate ways of reusing materials and objects in daily activities.
- Recognize that objects made of certain materials can be recycled.
- Identify, through observation, the same material in different objects.
- Compare objects constructed for similar purposes and identify the similarities and differences between their corresponding parts and the materials from which they are made.
- Identify materials commonly used in manufactured objects as well as the source of those materials.

UNIT 3: ENERGY IN OUR LIVES

- Recognize that the sun is the principal source of energy used on the surface of the earth.
- Identify food as a source of energy for themselves and other living things.
- Identify everyday uses of energy.
- Describe how our senses of touch, hearing, and sight, help us to control energy-using devices in the home, school, and community.
- Construct a manually controlled device that performs a specific task.
- Operate a simple device or system and identify the input and output.
- Ask questions about and identify needs and problems related to energy production or use in the immediate environment, and explore possible answers and solutions.
- Plan investigations to answer some of these questions or solve some of these problems.

- Use appropriate vocabulary in describing their investigations, explorations, and observations.
- Record relevant observations, findings, and measurements using written language, drawings, concrete materials, and charts.
- Communicate the procedures and results of investigations and explorations for specific purposes, using demonstrations, drawings, and oral and written descriptions.
- Describe the different forms of energy used in a variety of everyday devices.
- Identify everyday devices that are controlled manually.
- Identify devices they use that consume energy and list things they can do to reduce energy consumption.
- Select one of the most common forms of energy used every day and predict the effect on their lives if it were no longer available.

UNIT 4: EVERYDAY STRUCTURES

- Explain the function of different structures.
- Identify ways in which various structures are similar to and different from others in form and function.
- Classify various structures in their environment according to specific features and functions.
- Identify geometric shapes in ordinary structures.
- Describe patterns that are produced by the repetition of specific shapes or motifs in various materials and objects.
- Design and make different structures using concrete materials, and explain the function of the structure.
- Ask questions about and identify needs or problems related to structures in their immediate environment, and explore possible answers and solutions.
- Plan investigations to answer some of these questions or solve some of these problems.

▶

- Use appropriate vocabulary in describing their investigations, explorations, and observations.
- Record relevant observations, findings, and measurements, using written language, drawings, charts, and concrete materials.
- Communicate the procedures and results of investigations and explorations for specific purposes, using demonstrations, drawings and oral and written descriptions.
- Use appropriate natural and manufactured materials to make structures.
- Select appropriate tools and utensils.
- Use tools appropriately when joining and shaping various materials.
- Distinguish between structures and devices made by humans and structures found in nature.
- Explain the function of a structure that they have made and describe how they made it.
- Identify structures whose function is indicated by their shape.
- Examine different kinds of fasteners and indicate where they are used.
- Use and recognize the effects of different kinds of finishing techniques and processes on structures they have designed and made.
- Recognize that a product is manufactured to meet a need.
- Identify the action (input) required to operate an everyday system, and identify the response (output) of that system.
- Describe, using their own experience, how the parts of some systems work together.

UNIT 5: DAILY AND SEASONAL CYCLES

- Identify the sun as a source of heat and light.
- Compare the different characteristics of the four seasons.
- Use units of time related to the earth's cycles.

- Describe, using their observations, changes in heat and light from the sun over a period of time.
- Design and construct models of structures that would provide protection against local weather conditions.
- Ask questions about and identify needs or problems arising from observable events in the environment, and explore possible answers and solutions.
- Plan investigations to answer some of these questions or solve some of these problems.
- Use appropriate vocabulary in describing their explorations, investigations, and observations.
- Record relevant observations, findings, and measurements, using written language, drawings, concrete materials, and charts.
- Communicate the procedures and results of explorations and investigations for specific purposes, using demonstrations, drawings, and oral and written descriptions.
- Identify outdoor human activities that are based on the seasons and examine some of the solutions humans have found to make it possible to engage in these activities out of season.
- Identify characteristics of clothing worn in different seasons and make appropriate decisions about clothing for different environmental conditions.
- Identify features of houses that help keep us sheltered and comfortable throughout daily and seasonal cycles.
- Describe changes in the characteristics and behaviour of living things that occur on a daily basis.
- Describe changes in the characteristics, behaviour, and location of living things that occur in seasonal cycles.
- Describe ways in which humans modify their behaviour to adapt to changes in temperature and sunlight during the day.

▶

4

Program Principles

1. Effective science programs involve hands-on inquiry, problem solving, and decision making.

2. The development of students' skills, attitudes, knowledge, and understanding of STSE issues form the foundation of the science program.

3. Children have a natural curiosity about science and the world around them. This curiosity must be maintained, fostered, and enhanced through active learning.

4. Science activities must be meaningful, worthwhile, and relate to real-life experiences.

5. The teacher's role in science education is to facilitate activities and encourage critical thinking and reflection. Children learn best by doing, rather than by just listening. The teacher, therefore, should focus on formulating and asking questions rather than simply telling.

6. Science should be taught in correlation with other school subjects. Themes and topics of study should integrate ideas and skills from several core areas whenever possible.

7. The science program should encompass, and draw on, a wide range of educational resources, including literature, nonfiction research material, audio-visual resources, technology, as well as people and places in the local community.

8. Assessment of student learning in science should be designed to focus on performance and understanding, and should be conducted through meaningful assessment techniques carried on throughout the unit of study.

Program Implementation

Program Resources

Hands-On Science is arranged in a format that makes it easy for teachers to plan and implement.

Units are the selected topics of study for the grade level. The units relate directly to the learning expectations, which complement those established in the *Pan-Canadian Protocol* and *The Ontario Curriculum, Grades 1-8: Science and Technology, 1998* documents. The units are organized into several activities. Each unit also includes books for children, a list of annotated web sites, and references for teachers.

The introduction to each unit summarizes the general goals for the unit. The introduction provides background information for teachers, and a complete list of materials that will be required for the unit. Materials include classroom materials, equipment, visuals, and reading materials.

Each unit is organized into topics, based on the expectations. The topics are arranged in the following format:

Science Background Information for Teachers: Some topics provide teachers with the basic scientific knowledge they will need to present the activities. This information is offered in a clear, concise format, and focuses specifically on the topic of study.

Materials: A complete list of materials required to conduct the main activity is given. The quantity of materials required will depend on how you conduct activities. If students are working individually, you will need enough materials for each student. If students are working in groups, the materials required will be significantly reduced. Many of the identified items are for the teacher to use for display

▶

purposes, or for making charts for recording students' ideas. In some cases, visual materials – large pictures, sample charts, and diagrams – have been included with the activity to assist the teacher in presenting ideas and questions, and to encourage discussion. You may wish to reproduce these visuals, mount them on sturdy paper, and laminate them so they can be used for years to come.

Activity: This section details a step-by-step procedure, including higher-level questioning techniques, and suggestions, for encouraging exploration and investigation.

Activity Sheet: The reproducible activity sheets are designed to correlate with the expectations of the activity. Often, the activity sheets are to be used *during* the activity to record results of investigations. At other times, the sheets are to be used as a *follow-up* to the activities. Students may work independently on the sheets, in small groups, or you may choose to read through the sheets together and complete them in a large-group setting. Activity sheets can also be made into overheads or large experience charts. Since it is important for students to learn to construct their own charts and recording formats, you may want to use the activity sheets as examples of ways to record and communicate ideas about an activity. Students can then create their own sheets rather than use the ones provided.

Note: Activity sheets are meant to be used only in conjunction with, or as a follow-up to, the hands-on activities. The activity sheets are not intended to be the science lesson itself or the sole assessment for the lesson.

Extension: Included are optional activities to extend, enrich, and reinforce the expectations.

Activity Centre: Included are independent student activities that focus on the expectations.

Assessment Suggestions: Often, suggestions are made for assessing student learning. These assessment strategies focus specifically on the expectations of a particular activity topic (assessment is dealt with in detail on page 13). Keep in mind that the suggestions made within the activities are merely ideas to consider – you may use your own assessment techniques, or refer to the other assessment strategies on pages 15-25.

Classroom Environment

The classroom setting is an important aspect of any learning process. An active environment, one that gently hums with the purposeful conversations and activities of students, indicates that meaningful learning is taking place. When studying a specific topic, you should display related objects and materials, student work, pictures and posters, graphs and charts made during activities, and summary charts of important concepts taught and learned. An active environment reinforces concepts and skills that have been stressed during science activities.

Time Lines

No two groups of students will cover topics and material at the same rate. Planning the duration of units is the responsibility of the teacher. In some cases, the activities will not be completed during one block of time and will have to be carried over. In other cases, students may be especially interested in one topic and may want to expand upon it. The individual needs of the class should be considered; there are no strict time lines involved in *Hands-On Science*. It is important,

▶

however, to spend time on every unit in the program so that students focus on all of the curriculum expectations established for their grade level.

Classroom Management

Although hands-on activities are emphasized throughout this program, the manner in which these experiences are handled is up to you. In some cases, you may have all students manipulating materials individually; in others, you may choose to use small-group settings. Small groups encourage the development of social skills, enable all students to be active in the learning process, and mean less cost in terms of materials and equipment.

Occasionally, especially when safety concerns are an issue, you may decide to demonstrate an activity, while still encouraging as much student interaction as possible. Again, classroom management is up to you, since it is the teacher who ultimately determines how the students in his or her care function best in the learning environment.

Science Skills: Guidelines for Teachers

While involved in the activities of *Hands-On Science*, students will use a variety of skills as they answer questions, solve problems, and make decisions. These skills are not unique to science, but they are integral to students' acquisition of scientific literacy. The skills include initiating and planning, performing and recording, analyzing and interpreting, as well as communicating and the ability to work in teams. In the early years, basic skills should focus on science inquiry. Although the wide variety of skills are not all presented here, the following guidelines provide a framework to use to encourage students' skill development in specific areas.

Observing

Students learn to perceive characteristics and changes through the use of all five senses. Students are encouraged to use sight, smell, touch, hearing, and taste (when safe) to gain information about objects and events. Observations may be qualitative (by properties such as texture or colour), or quantitative (such as size or number), or both. Observing includes:

- gaining information through the senses
- identifying similarities and differences, and making comparisons
- sequencing events or objects

Exploring

Students need ample opportunities to manipulate materials and equipment in order to discover and learn new ideas and concepts. During exploration, students need to be encouraged to use all of their senses and observation skills. Oral discussion is also an integral component of exploration; it allows students to communicate their discoveries.

Classifying

This skill is used to group or sort objects and events. Classification is based on observable properties. For example, objects can be classified into living and nonliving groups, or into groups according to colour, shape, or size. One of the strategies used for sorting involves the use of Venn diagrams (either a double Venn or a triple Venn). Venn diagrams can involve distinct groups, or can intersect to show similar characteristics.

Venn Diagram With Distinctive Groups:

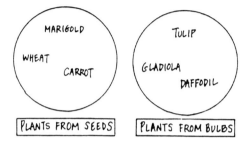

Intersecting Venn Diagram:

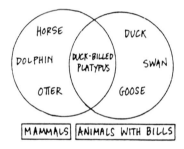

Measuring

This is a process of discovering the dimensions or quantity of objects or events and usually involves the use of standards of length, area, mass, volume, capacity, temperature, time, and speed. Measuring skills also include the ability to choose appropriate measuring devices, and using proper terms for direction and position.

In the early years, measuring activities first involve the use of *nonstandard* units of measure, such as unifix cubes or paper clips to determine length. This is a critical preface to measuring with standard units. Once standard units are introduced, the metric system is the foundation of measuring activities. Teachers should be familiar with, and regularly use, these basic measurement units.

An essential skill of measurement is *estimating*. Regularly, students should be encouraged to estimate before they measure, whether it be in nonstandard or standard units. Estimation allows students opportunities to take risks, use background knowledge, and learn from the process.

Length: Length is measured in metres, portions of a metre, or multiples of a metre. The most commonly used units are:

- millimetre (mm): about the thickness of a paper match
- centimetre (cm): about the width of your index fingernail
- metre (m): about the length of a man's stride
- kilometre (km): 1000 metres

Mass: Mass, or weight, is measured in grams, portions of a gram, or multiples of a gram. The most commonly used units are:

- gram (g): about the weight of a paper clip
- kilogram (kg): a cordless telephone weighs about 2 kilograms
- tonne (t): about the weight of a compact car

Note: When measuring to determine the heaviness of an object, the term *mass* is more scientifically accurate than the term *weight*. However, it is still acceptable to use the terms interchangeably in order for students to begin understanding the vocabulary of science.

Capacity: Capacity refers to the amount of fluid a container holds, and is measured in litres, portions of a litre, and multiples of a litre. The most commonly used units are:

- millilitre (ml): a soup spoon holds about 15 millilitres
- litre (l): milk comes in litre containers, or portions and multiples of a litre

Volume: Volume refers to the amount of space taken up by an object and is measured in cubic units, generally cubic centimetres (cm^3) and cubic metres (m^3).

▶

Note: Volume and capacity are often used interchangeably. However, a teacher should use the terms correctly in context, referring to liquid measure as capacity and space taken up as volume. Early years students are not yet expected to master the differences in the concepts and terminology, and can, therefore, be allowed to use the terms *volume* and *capacity* interchangeably.

Area: Area is measured in square centimetres, or portions and multiples thereof. By becoming familiar with the units of length, the teacher can understand area measurements by thinking of that unit in a two-dimensional form, such as square centimetres (cm²) and square metres (m²).

Temperature: Temperature is measured in degrees Celsius (°C). 21°C is normal room temperature; water freezes at 0°C and boils at 100°C.

Communicating

In science, one communicates by means of diagrams, graphs, charts, maps, models, symbols, as well as with written and spoken languages. Communicating includes:

- reading and interpreting data from tables and charts
- making tables and charts
- reading and interpreting data from graphs
- making graphs
- making labelled diagrams
- making models
- using oral and written language

When presenting students with charts and graphs, or when students make their own as part of a specific activity, there are guidelines that should be followed.

- A *pictograph* has a title and information on one axis that denotes the items being compared. There is generally no graduated scale or heading for the axis representing numerical values.

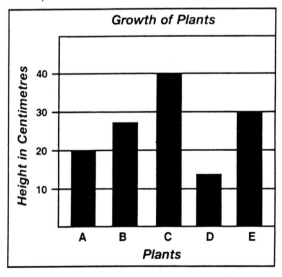

- A *bar graph* is another common form of scientific communication. Bar graphs should always be titled so that the information communicated is easily understood. These titles should be capitalized in the same manner as one

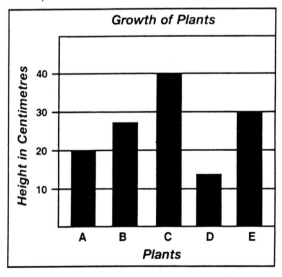

would title a story. Both axes of the graph should also be titled and capitalized in the same way. In most cases, graduated markings are noted on one axis and the objects or events being compared are noted on the other. On a bar graph, the bars must be separate, as each bar represents a distinct piece of data.

■ *Charts* also require appropriate titles, and both columns and rows need specific headings. Again, all of these titles and headings require capitalization as in titles of a story. In some cases, pictures can be used to make the chart easier for young students to understand. Charts can be made in the form of checklists or can include room for additional written information and data.

Objects in Water		
Object	Float	Sink
		✓
	✓	

Flowers We Saw		
Type	Colour	Diagram
daffodil	yellow	
rose	red	
lilac	purple	

Measuring Length		
Object	Estimate (cm)	Length (cm)
book	30 cm	27 cm
pencil	10 cm	16 cm

Communicating also involves using the language and terminology of science. Students should be encouraged to use the appropriate vocabulary related to their investigations, for example, *objects, material, solid, liquid, gas, condensation, evaporation, magnetic, sound waves,* and *vibration*. The language of science also includes terms like *predict, infer, estimate, measure, experiment,* and *hypothesize*. Teachers should use this vocabulary regularly throughout all activities, and encourage their students to do the same. As students become proficient at reading and writing, they can also be encouraged to use the vocabulary and terminology in written form. Consider developing whole-class or individual glossaries whereby students can record the terms learned and define them in their own words.

Predicting

A prediction refers to the question: What do you think will happen? For example, when a balloon is blown up, ask students to predict what they think will happen when the balloon is placed in a basin of water. It is important to provide opportunities for students to make predictions and for them to feel safe doing so.

▶

Inferring

When students are asked to make an inference, it generally means that they are being asked to explain why something occurs. For example, after placing an inflated balloon in a basin of water, ask students to infer why the balloon floats. Again, it is important to encourage students to take risks when making such inferences. Before explaining scientific phenomena to students, they should be given opportunities to infer for themselves.

Investigating and Experimenting

When investigations and experiments are done in the classroom, planning and recording the process and the results are essential. There are standard guidelines for writing up experiments that can be used with even young students.

- purpose: what we want to find out
- hypothesis: what we think will happen
- materials: what we used
- method: what we did
- results: what we observed
- conclusion: what we found out
- application: how we can use what we learned

Researching

Even at a young age, students can begin to research topics studied in class if they are provided with support and guidelines. Research involves finding, organizing, and presenting information. For best results, teachers should always provide a structure for the research, indicating questions to be answered, as well as a format for conducting the research. Suggestions for research guidelines are presented regularly throughout *Hands-On Science.*

Using the Design Process

Throughout *Hands-On Science,* students are given opportunities to use the design process to design and construct objects. There are specific steps in the design process:

1. Identify a need.
2. Create a plan.
3. Develop a product.
4. Communicate the results.

The design process also involves research and experimentation.

Do you or any of your students have a science question you want answered? E-mail your question to Randy Cielen, one of the authors of *Hands-On Science* and a member of the Science Teachers' Association of Manitoba and the National Science Teachers' Association. The address is: books@peguis.com.

Assessment

The *Hands-On Science* Assessment Plan

Hands-On Science provides a variety of assessment tools that enable teachers to build a comprehensive and authentic daily assessment plan for students.

Embedded Assessment

Assess students as they work, by using the questions provided with each activity. These questions promote higher-level thinking skills, active inquiry, problem solving, and decision making. Anecdotal records and observations are examples of embedded assessment:

- anecdotal records: Recording observations during science activities is critical in having an authentic view of a young student's progress. The anecdotal record sheet presented on page 15 provides the teacher with a format for recording individual or group observations.

- individual student observations: During those activities when a teacher wishes to focus more on individual students, individual student observations sheets may be used (page 16). This black line master provides more space for comments and is especially useful during conferencing, interviews, or individual student presentations.

Science Journals

Have the students reflect on their science investigations through the use of science journals. Several specific samples for journalling are included with activities throughout *Hands-On Science*. Teachers can also use notebooks or the black line master provided on page 17 to encourage students to explain what they did in science, what they

learned, what they would like to learn, and how they would illustrate their ideas.

Performance Assessment

Performance assessment is a planned, systematic observation and is based on students actually doing a specific science activity.

- rubrics: To assess students' performance on a specific task, rubrics are used in *Hands-On Science* to standardize and streamline scoring. A sample rubric and a black line master for teacher use are included on pages 18 and 19. For any specific activity, the teacher selects five criteria that relate directly to the expectations of students for the specific activity being assessed. Students are then given a check mark point for each criterion accomplished, to determine a rubric score for the assessment from a total of five marks. These rubric scores can then be transferred to the rubric class record on page 20.

Cooperative Skills

In order to assess students' ability to work effectively in a group, teachers must observe the interaction within these groups. A cooperative skills teacher assessment sheet is included on page 21 for teachers to use while conducting such observations.

Student Self-Assessment

It is important to encourage students to reflect on their own learning in science. For this purpose, teachers will find included a student self-assessment sheet on page 22, as well as a cooperative skills self-assessment sheet on page 23. Of course, students will also reflect on their own learning during class discussions and especially through writing in their science journals.

▶

Science Portfolios

Select, with student input, work to include in a science portfolio. This can include activity sheets, research projects, photographs of projects, as well as other written material. Use the portfolio to reflect the student's growth in scientific literacy over the school year. Black line masters are included to organize the portfolio (science portfolio table of contents on page 24 and the science portfolio entry record on page 25).

Note: In each unit of *Hands-On Science*, suggestions for assessment are provided for several lessons. It is important to keep in mind that these are merely suggestions. Teachers are encouraged to use the assessment strategies presented here in a wide variety of ways, and their own valuable experience as educators.

Date: _____

Anecdotal Record

Purpose of Observation: _____

Student/Group	Student/Group
Comments	**Comments**
Student/Group	**Student/Group**
Comments	**Comments**
Student/Group	**Student/Group**
Comments	**Comments**

Date: _____

Individual Student Observations

Purpose of Observation: _____

Student _____
Observations
Student _____
Observations
Student _____
Observations

Science Journal

Date: _____   Name: _____

[blank box]

Today in science I _____

_____

I learned _____

_____

I would like to learn more about _____

_____

Science Journal

Date: _____   Name: _____

[blank box]

Today in science I _____

_____

I learned _____

_____

I would like to learn more about _____

_____

Sample Rubric

Science Activity: _Looking at Seeds_

Science Unit: _____

Date: _____

| 5 – Full Accomplishment |
| 4 – Good Accomplishment |
| 3 – Substantial Accomplishment |
| 2 – Partial Accomplishment |
| 1 – Little Accomplishment |

| Student | Criteria | | | | | Rubric Score /5 |
	Follows Directions	Displays Curiosity	Makes Detailed Observations	Sorts and Classifies Seeds	Uses Appropriate Vocabulary to Communicate Ideas	
Jarod	✓	—	✓	✓	—	3
Ash	✓	✓	✓	✓	✓	5

SAMPLE

Rubric

Science Activity: _____

Science Unit: _____

Date: _____

5 – Full Accomplishment
4 – Good Accomplishment
3 – Substantial Accomplishment
2 – Partial Accomplishment
1 – Little Accomplishment

Student	Criteria						Rubric Score /5

Teacher: _____

Rubric Class Record

Student	Unit/Activity/Date									
	Rubric Scores /5									

Cooperative Skills
Teacher Assessment

Date: _____

Task: _____

_____

Group Member	Cooperative Skills					Rubric Score /5
	Contributes ideas and questions	Respects and accepts contributions of others	Negotiates roles and responsibilities of each group member	Remains focused and encourages others to stay on task	Completes individual commitment to the group	

Comments: _____

_____

_____

_____

_____

5 – Full Accomplishment
4 – Good Accomplishment
3 – Substantial Accomplishment
2 – Partial Accomplishment
1 – Little Accomplishment

Date: _____   **Name:** _____

Student Self-Assessment

Looking at My Science Learning

1. What I did in science: _____

_____

_____

2. In science I learned: _____

_____

_____

3. I did very well at: _____

_____

_____

4. I would like to learn more about: _____

_____

_____

5. One thing I like about science is: _____

_____

_____

Note: The student may complete this self-assessment or the teacher can scribe for the student.

Date: _____   Name: _____

Cooperative Skills Self-Assessment

Students in my group:

_____   _____

_____   _____

Group Work – How Did I Do Today?

Group Work	How I Did (✔)		
	🙂	😐	☹️
I shared ideas.			
I listened to others.			
I asked questions.			
I encouraged others.			
I helped with the work.			
I stayed on task.			

I did very well in _____

_____

Next time I would like to do better in _____

_____

Date: _____   Name: _____

Science Portfolio Table of Contents

Entry	Date	Selection
1.	_____	_____
2.	_____	_____
3.	_____	_____
4.	_____	_____
5.	_____	_____
6.	_____	_____
7.	_____	_____
8.	_____	_____
9.	_____	_____
10.	_____	_____
11.	_____	_____
12.	_____	_____
13.	_____	_____
14.	_____	_____
15.	_____	_____
16.	_____	_____
17.	_____	_____
18.	_____	_____
19.	_____	_____
20.	_____	_____

Date: _____ Name: _____

Science Portfolio Entry Record

This work was chosen by _____

This work is _____

_____

I chose this work because_____

_____

_____

_____

_____

Note: The student may complete this form or the teacher can scribe for the student.

- ✂ - - -

Date: _____ Name: _____

Science Portfolio Entry Record

This work was chosen by _____

This work is _____

_____

I chose this work because_____

_____

_____

_____

Note: The student may complete this form or the teacher can scribe for the student.

Unit 1

Characteristics and Needs of Living Things

Books for Children

Bergman, Thomas. *Seeing In Special Ways.* Milwaukee: Gareth Stevens Children's Books, 1989.

_____. *Finding a Common Language.* Milwaukee: Gareth Stevens Children's Books, 1989.

Bourgeois, Paulette. *Big Sarah's Little Boots.* Toronto: Kids Can Press, 1987.

_____. *Too Many Chickens.* Toronto: Kids Can Press, 1997.

Carle, Eric. *The Mixed-Up Chameleon.* New York: HarperFestival, 1998.

_____. *The Very Hungry Caterpillar.* New York: Philomel Books, 1994.

Cole, Joanna. *The Magic Schoolbus: Inside the Human Body.* New York: Scholastic, 1989.

Garner, Alan. *Jack and the Beanstalk.* New York: Doubleday, 1992.

Graves, Kimberlee. *Is It Alive?* Learn to Read Series, Level 1. Cypress, CA: Creative Teaching Press, 1994.

_____. *I Can't Sleep.* Learn to Read Science Series, Level 2. Cypress, CA: Creative Teaching Press, 1994.

_____. *What's In My Pocket?* Learn to Read Science Series, Level 2. Cypress, CA: Creative Teaching Press, 1994.

Graves, Kimberlee, and Rozanne Lanczak Williams. *Where Are You Going?* Learn to Read Science Series, Level 2. Cypress, CA: Creative Teaching Press, 1994.

Hallinan, P.K. *For the Love of Our Earth.* Nashville, TN: Ideals Publications, 1992.

Hutchins, Pat. *You'll Soon Grow Into Them, Titch.* New York: Greenwillow Books, 1983.

Intrater, Roberta Grobel. *Two Eyes, a Nose and a Mouth.* New York: Scholastic, 1995.

Kelley, Dr. Alden. *It Started As a Seed.* Learn to Read Series, Level 3. Cypress, CA: Creative Teaching Press, 1994.

MacKinnon, Debbie. *All About Me.* Hauppaugo: Barron's, 1994.

Martin, Bill Jr., and John Archambault. *Here Are My Hands.* New York: Holt, 1998.

Maynard, Christopher. *Why Are Pineapples Prickly?* Toronto: Scholastic, 1997.

McCue, Lisa. *The Little Chick.* New York: Random House, 1986.

Miller, Margaret. *My Five Senses.* New York: Simon & Schuster, 1994.

Peet, Bill. *The Wing-Ding-Dilly.* Boston: Houghton Mifflin, 1970.

Peterson, Jeanne Whitehouse. *I Have a Sister, My Sister Is Deaf.* New York: Harper & Row, 1977.

Pluckrose, Henry. *Exploring Our Senses – Hearing.* Milwaukee: Gareth Stevens Publishing, 1995.

_____. *Exploring Our Senses – Seeing.* Milwaukee: Gareth Stevens Publishing, 1995.

_____. *Exploring Our Senses – Smelling.* Milwaukee: Gareth Stevens Publishing, 1995.

_____. *Exploring Our Senses – Tasting.* Milwaukee: Gareth Stevens Publishing, 1995.

_____. *Exploring Our Senses – Touching.* Milwaukee: Gareth Stevens Publishing, 1995.

Saksie, Judy. *The Seed Song.* Learn to Read Series, Level 2. Cypress, CA: Creative Teaching Press, 1994.

Schwartz, David M. *Animal Feet.* Look Once, Look Again Plants and Animals Science Series. Cypress, CA: Creative Teaching Press, 1998.

▶

_____. *Animal Ears.* Look Once, Look Again Plants and Animals Science Series. Cypress, CA: Creative Teaching Press, 1998.

_____. *Animal Noses.* Look Once, Look Again Plants and Animals Science Series. Cypress, CA: Creative Teaching Press, 1998.

_____. *Animal Tails.* Look Once, Look Again Plants and Animals Science Series. Cypress, CA: Creative Teaching Press, 1998.

_____. *Animal Eyes.* Look Once, Look Again Plants and Animals Science Series. Cypress, CA: Creative Teaching Press, 1998.

_____. *Animal Mouths.* Look Once, Look Again Plants and Animals Science Series. Cypress, CA: Creative Teaching Press, 1998.

_____. *Animal Skin & Scales.* Look Once, Look Again Plants and Animals Science Series. Cypress, CA: Creative Teaching Press, 1998.

_____. *Animal Feathers & Fur.* Look Once, Look Again Plants and Animals Science Series. Cypress, CA: Creative Teaching Press, 1998.

_____. *Plant Stems & Roots.* Look Once, Look Again Plants and Animals Science Series. Cypress, CA: Creative Teaching Press, 1998.

_____. *Plant Leaves.* Look Once, Look Again Plants and Animals Science Series. Cypress, CA: Creative Teaching Press, 1998.

_____. *Plant Blossoms.* Look Once, Look Again Plants and Animals Science Series. Cypress, CA: Creative Teaching Press, 1998.

_____. *Plant Fruits & Seeds.* Look Once, Look Again Plants and Animals Science Series. Cypress, CA: Creative Teaching Press, 1998.

_____. *Monarch Butterfly.* Life Cycles Science Series. Huntington Beach, CA: Creative Teaching Press, 1999.

_____. *Bean.* Life Cycles Science Series. Huntington Beach, CA: Creative Teaching Press, 1999.

_____. *Sunflower.* Life Cycles Science Series. Huntington Beach, CA: Creative Teaching Press, 1999.

_____. *Wood Frog.* Life Cycles Science Series. Huntington Beach, CA: Creative Teaching Press, 1999.

_____. *Ladybug.* Life Cycles Science Series. Huntington Beach, CA: Creative Teaching Press, 1999.

_____. *Chicken.* Life Cycles Science Series. Huntington Beach, CA: Creative Teaching Press, 1999.

_____. *Jumping Spider.* Life Cycles Science Series. Huntington Beach, CA: Creative Teaching Press, 1999.

_____. *Maple Tree.* Life Cycles Science Series. Huntington Beach, CA: Creative Teaching Press, 1999.

_____. *Green Snake.* Life Cycles Science Series. Huntington Beach, CA: Creative Teaching Press, 1999.

_____. *Hummingbird.* Life Cycles Science Series. Huntington Beach, CA: Creative Teaching Press, 1999.

_____. *Horse.* Life Cycles Science Series. Huntington Beach, CA: Creative Teaching Press, 1999.

_____. *Fighting Fish.* Life Cycles Science Series. Huntington Beach, CA: Creative Teaching Press, 1999.

Seuss, Dr. *The Foot Book.* New York: Random House, 1968.

Suhr, Mandy. *Hearing.* Minneapolis: Carolrhoda Books, 1994.

Walt Disney Productions. *Bambi Grows Up.* New York: Random House, 1979.

▶

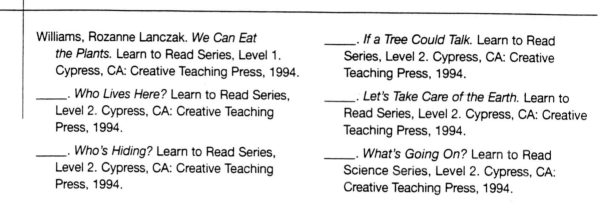

Williams, Rozanne Lanczak. *We Can Eat the Plants.* Learn to Read Series, Level 1. Cypress, CA: Creative Teaching Press, 1994.

_____. *Who Lives Here?* Learn to Read Series, Level 2. Cypress, CA: Creative Teaching Press, 1994.

_____. *Who's Hiding?* Learn to Read Series, Level 2. Cypress, CA: Creative Teaching Press, 1994.

_____. *If a Tree Could Talk.* Learn to Read Series, Level 2. Cypress, CA: Creative Teaching Press, 1994.

_____. *Let's Take Care of the Earth.* Learn to Read Series, Level 2. Cypress, CA: Creative Teaching Press, 1994.

_____. *What's Going On?* Learn to Read Science Series, Level 2. Cypress, CA: Creative Teaching Press, 1994.

Creative Teaching Press books are available from Peguis Publishers, Winnipeg

Web Sites

- **http://tqjunior.advanced.org/3750**

  An excellent introduction to the five senses. Click on Mr. Potato Head's eyes, nose, ears, mouth, or hands to learn more about each of the senses.

- **http://www.hhmi.org/senses**

  Web site for the Howard Hughes Medical Institute. Interesting and informative articles on seeing, hearing, and smelling the world.

- **http://weber.u.washington.edu/~chudler/neurok.html**

  Neuroscience for Kids has been created for teachers and students interested in the nervous system. Clicking on "Experiments and Activities" will take you to "The Senses." This site includes experiments, teacher resources, and extensive links.

- **http://www.fi.edu/tfi**

  The Franklin Institute Science Museum: scroll down and click on Minutes from ME, an index to classroom activities on hearing, smelling, and seeing.

- **http://ericir.syr.edu/Projects/Newton/11/tstesmll.html**

  Newton's Apple web site: effectively answers questions regarding the connections between taste and smell. Includes classroom activities.

- **http://family.com.go**

  Disney's online family magazine: enter "the senses" in their search field and find extensive articles and activities related to the senses.

- **http://www.cnib.ca/cci/braille/index.htm**

  Web site for the Canadian National Institute for the Blind. Learn the history of Braille and the Braille system.

- **http://dww.deafworldweb.org/**

  Learn American Sign Language with a complete interactive ASL dictionary. Includes articles written for and by deaf people. Link to Canadian Cultural Society for the Deaf.

- **http://innerbody.com/default.htm**

  Human anatomy online with interactive anatomy lessons.

- **http://sln.fi.edu/tfi/welcome.html**

  The Franklin Institute Science Museum: scroll down to view online exhibits of the heart, and to find quick science activities for the classroom.

- **http://kidshealth.org/**

  An excellent web site designed by medical professionals for students. Includes sections on the body, food and fitness, the senses, and safety issues.

- **http://www.defenders.org/eslc.html**

  Defenders of Wildlife web site contains high-resolution photographs and information on endangered species, as well as informative articles on biodiversity.

- **http://www.fi.edu/tfi/units/life/life.html**

  Linked to the Franklin Institute. An excellent site for researching "living things" with extensive cross-links and links to various resources. Find sites for classification, adaptation, ecosystems, biomes, habitats, the life cycle, survival, the food chain, and the energy cycle. ▶

- **http://www.seaworld.org/**

  Sea World web site: Animal Resources link is specifically designed to help you quickly find information about the animal kingdom. Sites include scientific classifications, ecology and conservation, and "fun facts."

- **http://eelink.net**

  Environmental Education on the Internet develops and organizes Internet resources to support, enhance, and extend effective environmental education.

- **http://romlx6.rom.on.ca/biodiversity**

  Discover the complex relationships among living things, from an ordinary Toronto backyard to three distinct ecosystems from Ontario and beyond – from "Hands-on Biodiversity" at the Royal Ontario Museum.

- **http://romlx6.rom.on.ca/ontario/fieldguides.html**

  Royal Ontario Museum: follow the easy steps described on these pages to design your own guide to common animals of your local area.

- **http://www.nature.ca/english/mnhome.htm**

  The Canadian Museum of Nature: explore Canada's natural history exhibits and collections. Includes "Ask a Scientist," and educational activities and games.

- **http://www.kortright.org**

  Located just ten minutes away from Metro Toronto, The Kortright Centre is Canada's largest environmental education centre – with over sixty different programs that have been designed to complement classroom curriculum.

Introduction

This unit focuses on the characteristics and basic needs of living things. Throughout this unit, students will demonstrate an understanding of the basic needs of animals and plants (e.g., the need for food, air, and water). They will investigate the characteristics of animals and plants. They will also demonstrate awareness that animals and plants depend on their environment to meet their basic needs, and they will describe the requirements for good health for humans.

Since it is not possible to bring all types of animals and plants into the classroom, you will need to collect pictures of a wide variety of living things. Involve the students in this project. Good sources for photographs and drawings are:

- old calendars
- magazines, especially *Ranger Rick, Owl, Chickadee, Highlights for Children,* and *National Geographic*
- local zoological, wildlife, humane, and naturalist societies

Contact organizations and agencies well in advance of the unit. You may be able to obtain materials and services such as booklets, posters, films, field-trip opportunities, and presentations for classroom use.

Put together a collection of fiction and nonfiction books about living things. Keep these in a separate part of your classroom library or science activity centre. Refer to them during activity, research, and free time.

This unit also incorporates concepts and activities that focus on the five senses. Through hands-on activities, students will use their five senses to describe the properties of objects and classify these objects. Students will also investigate the importance of the

senses and learn about ways that they can protect the body and preserve the senses.

Students will be expected to taste, smell, and touch various foods and objects. Stress the importance of safety throughout the unit: be aware of students' allergies and ensure that students understand that they are not to taste, smell, or touch objects without your permission. These issues can also be addressed through integration with health safety lessons.

Science Vocabulary

Throughout this unit, teachers should use, and encourage students to use, vocabulary such as: *body parts, muscle, heart, lungs, stomach, brain, bones, human, food groups, dairy, cereals, grains, sight, hearing, touch, taste, smell, characteristics, needs, living things, reproduce, animal, plant,* and *offspring.*

Materials Required for the Unit

Classroom: large sheets of paper (for tracing students' bodies), chart paper, large graph paper, drawing paper, mural paper, index cards, scissors, pencils, crayons, paints, paintbrushes, felt markers, coloured markers, glue, tape, clipboards, paper clips, unifix cubes, Plasticine, book or weights (approximately 2 kg)

Books, Pictures, and Illustrations: body-part labels (included), picture cards of internal body parts (included), internal body-part labels (included), pictures of animals at different growth stages, pictures of environment (included), magazines, pictures of living things, pictures of nonliving things, pictures of humans at different growth stages, pictures of animals and their offspring (included), pictures of plants, black line master of eyes (included), pictures of food, *I Have a Sister, My Sister Is Deaf* (a book by Jeanne

▶

Whitehouse Peterson), *Big Sarah's Little Boots*
(a book by Paulette Bourgeois), any version of
The Ugly Duckling, *Bambi Grows Up* (a Walt
Disney Production), *The Very Hungry
Caterpillar* (a book by Eric Carle), *The Little
Chick* (a book by Lisa McCue)

Household: adult's and children's outgrown
shoes, plastic knife, coloured wool (blue,
green, brown), plate, tray, salt, sport socks,
old dress shirts with buttons

Equipment: stethoscope

Other: parts of plants (e.g., pine cones,
leaves, sunflower seeds, oranges), balloon,
mirrors, live plants of various species, apple,
skeleton or bone sample, actual living things
(e.g., plants, hamster, fish), Hula-Hoops,
samples of different foods, popcorn, air
popcorn popper, paper cups, paper plates,
napkins, large empty box, blindfolds, earplugs

1 | Body Outline

Materials

- large sheets of paper for tracing students' bodies (one sheet per student)
- cut-out labels of different body parts (included) (1.1.1). (Make a copy for each student.)
- scissors
- pencils, crayons
- paint, paintbrushes
- glue

Activity

Begin by playing games that involve naming and pointing to different parts of the body (e.g., Simon Says or Hokey-Pokey).

Trace each student's body onto large sheets of paper (or have them work in pairs to do this).

Cut out the shapes and add details (mouth, eyes, toes, fingers, and so on) to the body outline, using crayons or paints.

As a group, have students identify each body part, using one of the completed tracings as a model. Place the appropriate label (1.1.1) onto each body part and discuss its function.

Display the body outlines around the classroom for students to observe. It is also fun to "sit" the students' body outlines on their chairs for an open house or at parent-teacher interviews.

Activity Sheet

Directions to students:

Cut out the body-part labels and glue them onto the correct part of the body outline (1.1.2).

Extension

Make a class book called *My Incredible Body Parts* based on *Here Are My Hands* by Bill Martin Jr. and John Archambault. Assign each student one part of the body or, depending on the size of your class, have the students work in pairs. Have the students cut out pictures of their assigned body part from magazines and newspapers and glue the pictures onto their page in the big book. Remind them to leave a space at the bottom of the page for a sentence that tells something about that body part. Create a pattern when writing the sentences, using a format similar to *Here Are My Hands*. Keep the book in a science or reading corner of the classroom.

Read related books that can be used as patterns to make class books; for example, *The Foot Book* by Dr. Seuss.

Assessment Suggestion

Using their own body outlines or their activity sheet, have the students individually tell you the name and function(s) of each of the labelled body parts. Use the individual student observations sheet on page 16 for recording results.

Date: _____   **Name:** _____

Body Part Labels

leg 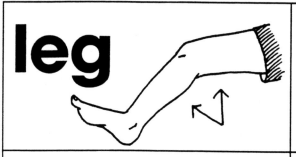	arm 
eye 	nose 
ear 	mouth 
knee	chest
foot 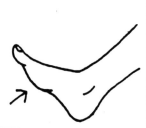	hand 

Date: _____   Name: _____

My Body

leg

eye

ear

arm

nose

mouth

knee

foot

chest

hand

2 | Inside the Human Body

Science Background Information for Teachers

Bones: The skeleton gives shape to and supports the body.

Heart: The heart pumps blood to all parts of the body.

Lungs: Air travels to our lungs so that we can breathe. Lungs expand as they are filled with air and contract as the air is pushed out.

Brain: The brain is the "boss of the body." It controls all of the functions of the body.

Stomach: Food travels from our mouth to our stomach through a tube called the esophagus. Acid in the stomach breaks down the food so that it can be used for energy in different parts of the body.

Muscles: Muscles tighten and relax to enable different parts of the body to move.

Note: These are complex concepts for young students. The focus should be on having students understand what a skeleton is, identify the body organs, and describe, in their own words, the function of each.

Materials

- apple
- plastic knife
- glue
- scissors
- labels of internal body parts (included) (1.2.1)
- picture cards of internal body parts (heart, brain, stomach, lungs, muscles, and bones) (included) (1.2.2-1.2.7)
- balloon
- stethoscope
- skeleton (or fabricated sample of a bone)
- book or weight (2 kg)
- student body outline example (see page 35)

Activity A

Have the students sit in a circle. Place an apple on the floor in the centre of the circle. Have the students describe the apple, brainstorming all of the things that they can see. Cut the apple in half and have the students describe the inside of the apple.

Explain that, like the apple, the inside of our body is different from the outside of our body.

Note: Since students generally are familiar with the outside and the inside of an apple, you may wish to introduce them to other fruits and vegetables (e.g., spaghetti squash, pomegranate). At this time, you may also want to mention that the skin of the apple protects the inside of the apple just like our skin and bones protect our inside body parts.

Ask the students what clues they can see or feel on the outside of the body that make them think there is something on the inside, too (e.g., I can feel my bones; I can feel my heart beat when I run; I bleed when I am cut).

Ask the students to name parts that are inside their body. Using a body outline from the previous lesson, label the parts as they are mentioned (1.2.1).

As each part is identified, show the large picture card of it (1.2.2-1.2.7). Discuss each part and describe its function in simple terms (e.g., my lungs help me breathe).

Activity Sheet

Directions to students:

Cut out the inside body parts. Paste each part where it belongs on the body outline (1.2.8).

Activity B

To provide students with hands-on experiences, conduct the following investigations into the functions of internal body parts:

2

The heart: Have students listen to their hearts beat, using the stethoscope. Have them run around the gym or recess field, then discuss what happens to their heart beats.

The lungs: Have students take a deep breath in and watch their chests to see what happens. Now have them let the air out and watch what happens. Have them blow up a balloon and let it deflate to model how lungs expand and contract.

The bones: Discuss which bones the students can feel in their bodies. Pass around the bone sample for students to feel, and discuss how the bones' strength gives form to the body.

The muscles: Have students put their left hand on their right arm muscle (bicep), and tighten the muscle to feel it contract. Now have them lift a weight (such as a book) to further show how the muscles contract when used.

Extensions

■ Access medical clinics, labs, and local hospitals and request old x-ray films. You can use these to show to students. X-ray films are especially effective when displayed on classroom windows. Students can use the x-rays to help identify their own bones.

■ For a further look at internal body parts and their functions, read *The Magic Schoolbus: Inside the Human Body,* by Joanna Cole.

Assessment Suggestion

Have each student identify parts inside the human body and explain the function of each. Use the rubric on page 19 to record results.

Labels for the Human Body

muscles

heart

lungs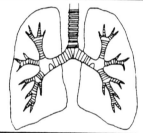

stomach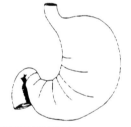

brain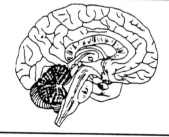

bones

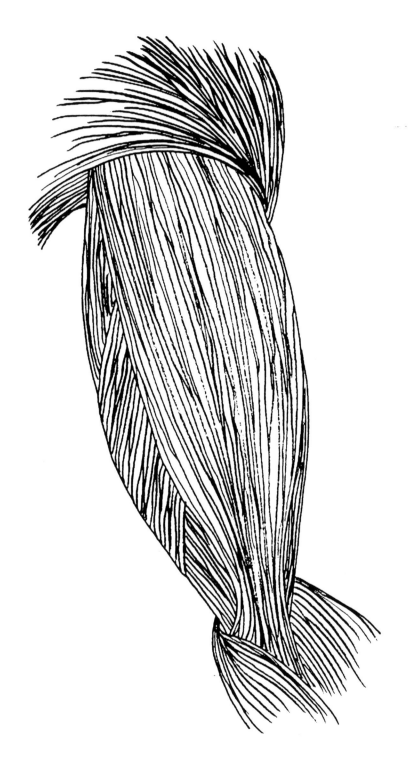

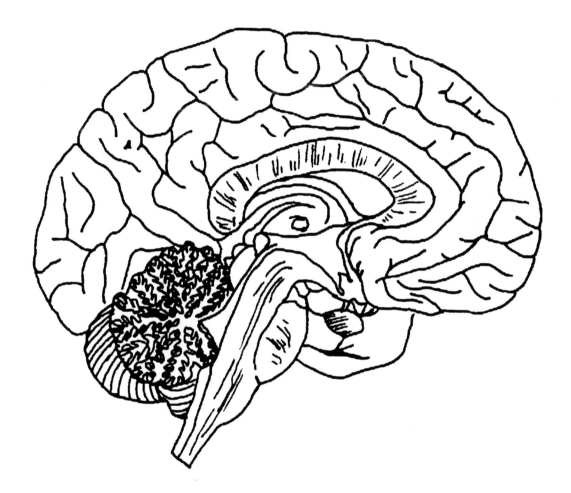

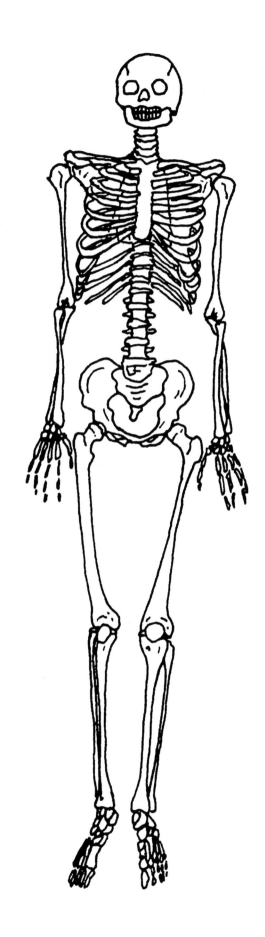

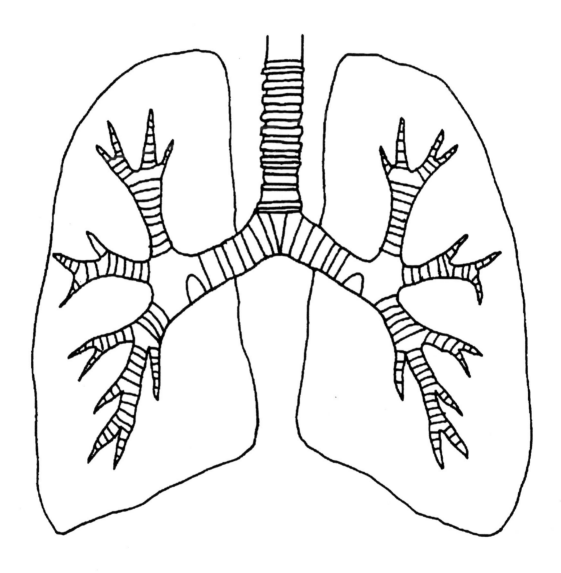

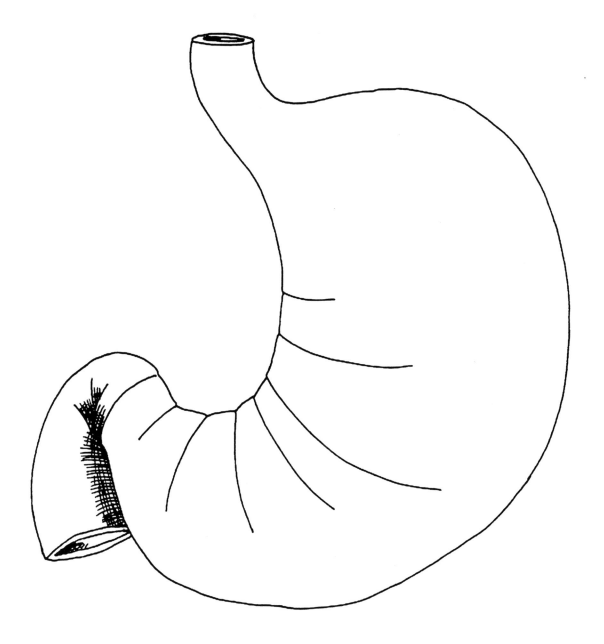

Inside My Body

stomach

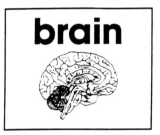

brain

muscles

heart

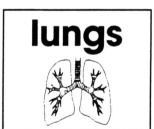

lungs

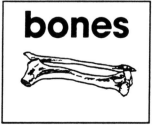

bones

3 Variations in Humans

Materials

- mirrors (one for every two students)
- glue
- large graph paper
- black line master of eyes for picture graph (cut these out beforehand) (included) (1.3.1)
- chart paper
- tape
- coloured markers, crayons
- index cards
- coloured wool (green, blue, and brown) for eye colours

Activity

Have students form a circle and observe one another's eyes. Have them describe the similarities and differences that they observe.

Once the topic of colour is raised, have the students describe the various eye colours of classmates. Pass out the mirrors so students have an opportunity to observe their own eyes as well.

Print the eye colours on index cards, using appropriate coloured markers. Take the wool and create a triple Venn diagram (make sure it is large enough for several students to stand inside). Use the index cards as labels for each circle. Have the students sort themselves on the Venn diagram according to eye colour and count the number of students in each group. Record the results on chart paper.

Note: The most common colours will be shown on the Venn diagram. Students with other eye colours will stand outside the circles.

Once all results have been recorded, compare the number of students with each eye colour. Discuss how some colours are more common than others.

Make a concrete graph by placing the eye colour index cards on the floor and having the students line up behind the appropriate label.

Now, using graph paper, create a picture graph with the cut-out eyes from the black line master (1.3.1). Have students print the colour of their eyes, then colour in the eyes and glue them on the graph. Use the index cards as labels for the horizontal axis.

Eye Colour — picture graph with vertical axis labelled *Number of Students* and horizontal axis labelled *Eye Colour* with columns Brown, Blue, Green, Grey, Hazel.

Discuss other similarities and differences in humans, such as hair colour and height. Emphasize the uniqueness of all the students by having them brainstorm reasons why they are special.

Divide the class into working pairs by having students find a partner who looks different from themselves. Each pair can then complete the activity sheet.

3

Activity Sheet

Directions to students:

Use the activity sheet to describe the colour of your eyes and hair, and then tell something special about yourself. Next describe the colour of your partner's eyes and hair, and tell something special about him or her. Draw pictures of your eyes and hair and of your partner's eyes and hair (1.3.2).

Activity Centre

Provide free-standing mirrors, along with drawing paper and crayons. Have the students observe their own features and create self-portraits.

Black Line Master for Picture Graph

My eyes are	My eyes are
_____	_____
Name _____	Name _____
My eyes are	My eyes are
_____	_____
Name _____	Name _____
My eyes are	My eyes are
_____	_____
Name _____	Name _____

Date: _____   **Name:** _____

My eyes are _____

My hair is _____

I am special because _____

_____

My friend's eyes are _____

My friend's hair is _____

My friend is special because _____

_____

4 | Growth of Humans

Materials

- pictures of humans at different growth stages (pictures of babies, students, and adults can be easily collected from department store catalogues)
- chart paper, felt pens
- students' outgrown shoes (Have each student bring one from home. The shoes should be significantly smaller than those they are currently wearing. If smaller shoes are not available, shoes worn by younger children/siblings can be used. The concept of human growth will still be evident in the activity.)
- one of your own outgrown shoes, and one you are currently wearing
- paper clips or unifix cubes
- *Big Sarah's Little Boots*, a book by Paulette Bourgeois

Activity

Make a random display of pictures of humans at different stages in their life. Ask the students:

- What do you notice about these pictures?
- How could you arrange these pictures in order? (e.g., youngest to oldest)
- Why would you arrange the pictures in this way?

Once the pictures are organized in order from youngest to oldest, ask the students:

- What happens to your bodies as you get older? (e.g., we grow, we lose baby teeth)
- In what ways are you different now from when you were a baby?

Make a list of their suggestions (e.g., lost baby teeth, wear bigger clothes, can dress self) on the chart paper.

Read aloud *Big Sarah's Little Boots* by Paulette Bourgeois to the students.

Ask students to display the outgrown shoes that they have brought from home. Display your own as well.

On chart paper, demonstrate how to trace around a shoe. Take your own shoes and trace around the outgrown one and the current one.

Have students estimate the length of both your traced shoes. Record the estimates on the chart paper.

Now use the paper clips or unifix cubes to demonstrate how to measure the lengths of the two shoes traced. Record the lengths on the chart paper and compare these measurements to the students' estimates.

Give the students some paper clips or unifix cubes. They will now have an opportunity to estimate and measure the length of their own shoes during the activity sheet task.

Activity Sheet

Note: Copy this activity sheet on longer paper so there is enough room to easily trace shoes.

Directions to students:

Trace around your outgrown shoe, then trace around one of the shoes you are wearing now. Estimate the length of each and record your estimates. Now measure the length of both and record the results on your activity sheet (1.4.1).

Extensions

- Challenge the students to draw a picture of the size they think their shoe might be a year from now. Have them measure the length.

▶

4

- Have students select five special events that have occurred in their lives. Ask them to draw a picture of each event and sequence them on a time line by attaching pictures to a string with clothespins. Hang the time lines in the classroom for all to see.

- Measure and compare the height of the students in September, January, and June.

- Read related books such as *You'll Soon Grow into Them, Titch* by Pat Hutchins.

- Have students bring baby pictures from home to display on a bulletin board. Also display the students' recent school pictures and, using string and tacks, have the students attempt to match baby pictures to recent pictures.

Assessment Suggestion

Through observation, determine if students can estimate, measure, and compare shoes of different sizes. Determine if they can successfully count the nonstandard units of measurement in order to draw comparisons. Use the rubric on page 19 to record results.

Date: _____   **Name:** _____

Here I Grow Again

Estimate _____   **Estimate** _____

Length _____   **Length** _____

5 Four Basic Food Groups

Science Background Information for Teachers

The Canadian Food Guide identifies the four food groups as:

Milk and Dairy
e.g., cheese, milk, yogurt, ice cream

Fruits and Vegetables
e.g., lettuce, apples, carrots, pineapple

Meat (and Alternatives)
e.g., beef, chicken, eggs

Breads and Cereals (Grains)
e.g., bagels, crackers, oatmeal

Materials

- 4 Hula-Hoops
- pictures of foods from the different food groups from magazines and flyers
- samples of different types of foods from the different food groups
- 4 food-group labels (print the name of each food group on index cards)
- tray
- scissors, glue

Activity

Have the students sit in a circle. Review the lesson on human growth (see pages 52-54). Ask the students what they think they need in order to grow. When the topic of food arises, put the food (or pictures of food) on a tray in the centre of the circle. Spread the four Hula-Hoops out around the food. Tell the students that you are going to sort the food by a rule, but you are not going to tell them what the rule is.

Have each Hula-Hoop represent one of the four food groups. Sort the food, one by one, into each of the hoops. Ask the students:

- What was my rule?
- How did you know?
- What group would everything in the first Hula-Hoop belong to?
- Where do these foods come from?

Repeat with each Hula-Hoop until all four groups have been named. Place the food-group labels in their respective Hula-Hoop.

Note: Students may not initially use the formal names for each of the food groups. When they have distinguished between each food group, introduce the formal names. You may also need to explain why eggs, beans, and nuts are in the Meats and Alternatives group. Do not worry about introducing the word *protein* to students at this age. Rather, explain that all of the foods in this group help our muscles grow strong.

Activity Sheet

Note: This activity sheet can be enlarged onto bigger paper to allow more room for pictures.

Directions to students:

Cut out pictures of foods from magazines and grocery store flyers. Sort the pictures into the four food groups and glue them onto your sheet (1.5.1).

Extension

Have students sort their lunch before they eat it. Ask:

- Do you have something from each food group?
- Do you have a healthy lunch?
- How could you have something from each of the four food groups but only eat one or two things? (e.g., ham, cheese, and lettuce sandwich)
- What other types of food could you put in your lunch that would make it more healthy?

5

Activity Centre

Set up a kitchen centre in the classroom, complete with dishes and food (plastic food, pictures of food, or food containers such as empty cereal boxes and milk jugs). Have students create healthy meals to serve to others at the centre.

Assessment Suggestion

Have students cut out pictures of foods to make a healthy breakfast, lunch, or supper. Have them glue these onto construction paper. Use the anecdotal record sheet on page 15 to record results.

The Food Groups

Sort your pictures into the four food groups and glue them into the correct box.

Fruits and Vegetables	Breads and Cereals
Milk and Dairy	Meats (and Alternatives)

6 The Senses

Materials

- popcorn
- air popcorn popper with transparent lid
- paper cups or paper plates
- napkins
- salt
- large empty box to cover popcorn popper
- pencils

Activity

Popping corn is one experience that involves all the senses. Prepare the popcorn popper while the students are out of the room. Hide the popcorn popper under a box on a low table or on the floor in the centre of the room. As the students enter, have them sit or stand around the box. Ask the students:

- What do you think is under the box?

Plug in the popcorn popper. Encourage the students to describe the sounds and smells coming from the box, and predict what is under it. Remove the box and allow the students to watch as the popcorn finishes popping. Sprinkle salt on the popcorn and give each student a small cup or plate of popcorn.

Ask the students:

- Is the popcorn warm? How do you know?
- What part of your body helps you to feel?
- What does popcorn look like now?
- How did the popcorn look before it popped?
- How has the popcorn changed?
- What part of your body helps you to see?
- What does popcorn taste like?
- What part of your body helps you to taste?
- Can you taste any salt?
- What did the popcorn sound like as it was coming out of the popper?
- What sound does the popcorn make when you eat it?

- What part of your body helps you to hear?
- What sense(s) did you use to discover what was under the box?
- What does the popcorn smell like?
- What part of your body helps you to smell?

Review the senses the students have used to learn about popcorn.

Note: Be sure to supervise the hot popcorn popper at all times.

Activity Sheet

Directions to students:

Draw a picture that shows something you like to do that uses each of your senses (e.g., I see with my eyes, I like to see the sunset) (1.6.1).

Extension

Discuss how your senses can affect your safety. Ask students:

- How do you sense danger? (Some examples are: smell smoke, hear a dog growling, see a car coming toward you as you are about to cross the street.)

Activity Centre

Create a senses centre.

Note: Before selecting items for the senses centres, be aware of any student allergies.

Touch: Make a "Feely Box." Cut a hole in the box large enough for a student's arm to fit through. Fill the box with various items and have the students identify the items without looking at the objects.

Smell: Fill old film containers with various things that have distinctive odours (e.g., perfume, vinegar, cinnamon, vanilla).

6

Note: If using a liquid such as perfume, soak a cottonball with the perfume, and place the cottonball in the film container.

Punch a couple of holes in the top of the lid. Have the students identify the items using their sense of smell. Remind students that when they are smelling objects they should not inhale the fumes directly, but should wave a hand over the object toward their nostrils.

Taste: Place a number of different foods on a tray. Have the students identify the foods. Blindfold the students and see if they can identify the foods through taste.

Hearing: Tape-record a variety of familiar sounds (e.g., school bell, voices, birds chirping). Have the students identify the sounds they hear.

Sight: Have the students work with a partner or in a small group. Place a number of familiar classroom objects on a tray (e.g., chalkboard eraser, clock, felt pen, crayon). Invite the students to identify each of the items on the tray. Have the students close their eyes while one student removes an object from the tray. Ask the students to open their eyes and identify the missing object.

My Five Senses

I see with my	I like to see
I hear with my	I like to hear
I smell with my	I like to smell
I taste with my	I like to taste
I feel with my	I like to feel

7 | Loss or Limitation of the Senses

Materials

- blindfolds
- earplugs
- sport socks
- old dress shirts with buttons
- *I Have a Sister, My Sister Is Deaf,* a book by Jeanne Whitehouse Peterson
- chart paper, felt pens

Activity

Begin by reading *I Have a Sister, My Sister Is Deaf.* Following the story, ask the students:

- What activities could the sister not do because she was deaf?
- There were some activities that she could do very well. What were they?
- Why do you think she was so good at these activities? What senses did she use?
- Have you ever lost the use of one of your senses (e.g., loss of smell or taste when you have a cold)?
- What was it like?

Have the students brainstorm other physical conditions that limit the use of one or more of the senses (e.g., blindness).

Provide the students with several experiences simulating the loss of a sense:

- loss of sight: Work in partners. Blindfold one partner. Have the other partner lead the blindfolded student around the classroom or gymnasium.
- loss of hearing: Have students wear earplugs around the classroom for a period of time.
- physical condition (limited use of part of the body, e.g., limb): Have students put on a dress shirt over top of their clothes, then place a sport sock over each of their hands. Have them attempt to do up the buttons on the shirt with their hands in the sport socks.

Following the activities, have the students share their experiences. Ask:

- What was it like? How did you feel?
- How did you cope with your physical condition? Did you rely on your other senses?

Refer back to your original list of physical conditions. Discuss ways that people can use aids to help them sense, and communicate with, the world around them (e.g., eyeglasses, seeing-eye dogs, wheelchairs, sign language, and hearing aids).

Ask the students what they can do to help someone with a physical condition.

Activity Sheet

Note: You may find it beneficial to invite a guest speaker with a physical condition to present to the class prior to having students complete the activity sheet. This will provide students with the background knowledge necessary to complete the task.

Directions to students:

You will each be responsible for making one page for our class book called *Helping Everyone Sense the World Around Them.* Draw a picture of how humans or some object can help those with a physical condition. Write a sentence to describe your picture (1.7.1).

Note: Students may use invented spelling to write their sentences, or you can scribe the sentences for your students.

Extension

Invite guest speakers with various physical conditions to the classroom. Have the guest speakers share their experiences.

7

Activity Centre

Provide samples of Braille for the students to touch. Display posters of the American Sign Language and encourage students to learn to sign the alphabet.

Assessment Suggestion

Have students orally present their page from the class book. Prior to presentations identify, as a class, five criteria for the activity. For example:

1. I made a detailed picture.
2. I showed an example of someone with a physical condition.
3. I showed an example of how to help someone with a physical condition.
4. I presented my page using a clear speaking voice.
5. I worked hard on my page.

Use the rubric on page 19 to list these criteria and record results.

Helping Everyone Sense
the World Around Them

8 | Characteristics of Living Things

Science Background Information for Teachers

The characteristics of living things are:

1. they need food, water, air
2. they grow
3. they reproduce
4. they die

Materials

- living things such as plants, a hamster, or fish, if possible
- pictures of nonliving things such as toys, cars, and household items
- pictures of living things, including humans, other animals, and plants
- chart paper, felt pens, scissors, glue
- drawing paper, crayons

Activity

Display a variety of living and nonliving things for students to observe, manipulate, and discuss. Ask:

- Which objects here are alive?
- How do you know they are alive?
- Which objects are not alive?
- How do you know they are not alive?

Have the students classify the objects into living and nonliving groups.

Remove the nonliving objects and have the students examine the living things. Challenge the students to sort them into groups such as:

- humans, other animals, and plants
- animals and plants
- things with legs and no legs
- things that move and do not move
- things with hair and no hair
- things with eyes and no eyes

Now encourage them to discuss what they know about living things. Ask:

- What is the same about all these objects? (e.g., they all grow)
- What does a living thing need to stay alive?
- Does a living thing always stay the same size?
- Will it live forever?
- What will happen to all living things some day?
- Where does a chicken come from?
- Where does a tree come from?
- What does this tell you about living things? (e.g., they all reproduce or have young)

These questions will encourage the students to infer, predict, and interpret what they observe. Introduce the word *reproduce* to the students and use the term often in subsequent lessons so that it becomes part of the students' vocabulary. Of course, it is also acceptable for the students to refer to reproduction of living things as "having babies" or "having young."

During the activity and subsequent discussion, record on chart paper students' responses as to the characteristics of living things.

Activity Sheet

Directions to students:

Choose one thing you know about all living things and print it on your activity sheet. Use the list on the chart paper for ideas. Now draw a picture to go with your sentence (1.8.1).

Note: You may wish, as a class, to describe examples prior to students working on their activity sheets. For example, to depict the reproduction of living things, students can draw pictures of a chicken and an egg or a tree and a seedling.

8

Extensions

- Play What Is My Rule?. Have students select "a rule" for sorting pictures of living things (e.g., sort a set of pictures of living things into a group with fur and a group without fur). Once the student has sorted the pictures based on the rule, ask another student to guess what rule was used to sort the pictures. Use the extension sheet for sorting (1.8.2).

- Visit a local zoo, bird sanctuary, nature centre, pet store, or wildlife reserve. Many of these offer winter programs as well, so do not limit yourself to spring field trips.

- Invite guest speakers from any of the above to visit the classroom with slide show presentations or live animals.

- Go on a nature hunt. Give each student two paper bags. Have the students label one bag Living Things and the other bag Nonliving Things. Students can then collect objects for each bag.

Note: Encourage students to collect items that come from living things (such as feathers, leaves, seeds), rather than collect live specimens.

Activity Centre

- Select several pictures of living things and cut them into puzzle pieces. Challenge students to identify the living thing by observing just one puzzle piece. Then have the students gather the other pieces for that living thing and put the puzzle together.

- Have students create various animals from Play-Doh or Plasticine. Encourage them to mix colours to get appropriate animal colours.

Assessment Suggestion

Using individual conferences or interviews, ask the student to identify common characteristics of living things. Provide pictures of several animals and ask the student to identify one way in which each animal is like a human (e.g., they both have legs, they both have ears, they both breathe, they both grow). Use the anecdotal record sheet on page 15 to record results.

Date: _____ Name: _____

Living Things

One thing I know about all living things is

_____

_____

_____

_____

Name: _____

What Is My Rule?

Date: _____

9 Using the Senses to Classify Living Things

Materials

- clipboards (Clipboards can also be made with a piece of cardboard and a clothespin. Use string to attach pencils to the clipboards.)
- pencils
- chart paper, felt pens

Note: Select an area near the school where students can use their senses to explore living things. Before doing this activity, check the area for poisonous plants and any other dangers.

Activity

Explain to students that they are going to go on a sensory walk around the school playground (or a nearby park). Ask:

- What are the names of each of the five senses?
- You will be using all of your senses on this walk, except for your sense of taste. Why will you not be using your sense of taste?
- As you walk, you are going to be looking for living things. What kinds of living things do you think you will see on your walk?
- Do you think you will smell any living things on the walk?
- Do you think you will hear any living things on the walk?
- What kinds of living things do you think you might be able to touch?

Review the activity sheet with the students. Explain that they are to record, using pictures or simple words, what they see, hear, touch, and smell.

Provide each student with a clipboard, pencil, and the activity sheet before going on the walk.

After returning from the walk, have students share some of the things they sensed on their walk.

Divide the chart paper into four sections: Things We Saw, Things We Heard, Things We Touched, Things We Smelled.

Record the students' experiences on the chart paper.

Activity Sheet

Note: The activity sheet is to be completed during the activity.

Directions to students:

Record things you see, smell, touch, and hear on your sensory walk. Use drawings or words to record your observations (1.9.1).

Extension

Have the students create their own sensory book. Encourage them to use their senses when they visit different places (e.g., supermarket, bakery, restaurant, theme park). Have students create a pattern book using the activity sheet; for example:

Using My Senses at the Fair

I smell French fries.

I hear music.

I see rides.

I touch sticky cotton candy.

Date: _____   **Name:** _____

Sensory Walk

I see...	I smell...
I hear...	**I touch...**

10 | Needs of Living Things

Materials

- various pictures of living things (from previous lessons)
- chart paper, felt pens

Activity A

Observe and discuss the needs of humans, focusing on what they need to stay alive. Ask the students:

- What does a human need to stay alive?
- How do humans breathe?
- What do humans eat?
- How do we make sure that we are eating healthy food?
- Where do we get our food from?
- Are there other ways for humans to get food other than from the grocery store?
- What else does a human need to survive? (e.g., shelter)

Divide chart paper into two columns. At the top of the first column, print the heading Needs of Humans. As students describe the needs of humans, list them on this side of the chart paper.

Activity B

Display the pictures of living things, other than humans, for the students to observe and discuss. Ask:

- What does an animal need to stay alive?
- What does a plant need to stay alive?
- How do different animals breathe? (e.g., lungs, gills)
- What do different animals eat?
- How do these animals get their food?
- What else do animals need to stay alive? (e.g., water)

- What else do plants need to stay alive? (e.g., water, sunlight)
- Where do different plants and animals get water? (e.g., lakes and streams)
- What do you think would happen to plants and animals if it never rained again and all the lakes and rivers dried up?

At the top of the second column on the chart paper, print the heading Needs of Other Living Things. As students describe the needs of other living things, list them on the chart paper.

Review each list and compare the basic needs of humans with the basic needs of other living things. Circle the needs that humans and other living things have in common.

Activity C

Have each student in the class select an animal that he or she would like to study further. Have them read picture books about these animals to identify the following:

- what the animal looks like
- how the animal moves
- what the animal eats
- where the animal lives

This information will be used to complete the activity sheet.

Note: As a class, you may first want to research one animal.

Activity Sheet

Directions to students:

Make a picture book showing what you have learned about your special animal. Finish each sentence and draw a picture to go with it (1.10.1).

10

Example:

My special animal, the robin.

The robin is a bird.

The robin is brown and red.

The robin moves by flapping its wings and flying.

The robin eats worms.

The robin lives in a nest in a tree.

Extension

Select a classroom pet that the students can look after. Stress the importance of proper care and handling of the pet. Review the needs of the pet prior to the arrival of the animal and list these on chart paper. As a class, discuss ways to ensure that everyone understands how to properly care for the pet.

Through reading about the animal, have the students plan a home for the pet, ensuring that its needs will be met.

Introduce the animal to the students. Choose a name for the animal. Have a duty roster so that all students experience the responsibility of caring for and looking after the needs of the animal. A duty roster also eliminates overfeeding the animal.

Set guidelines for visiting the animal centre, such as how many students can be there at the same time and for how long. Students should not be allowed to visit the centre without some form of supervision.

Note: A number of animals are suitable for the classroom environment, and do not take an unreasonable amount of time to care for and maintain. Fish, for example, are easily cared for, and a fish tank provides an excellent observation centre for students. Guinea pigs can be appropriate classroom pets; they are large enough for students to handle, are not fast enough to scurry away, and are known to be gentle and affectionate. You may also choose to have guest pets visit the classroom for a month at a time, such as a bird, a rabbit, a gerbil, a snake, and so on. In this way, a greater variety of animals can be introduced to students.

Note: Students' allergies must be considered when selecting a classroom animal.

My special animal, the _____.

The _____ is_____.

The _____ is_____.

The _____ moves by

_____.

The _____ **eats**_____.

The _____ **lives**

_____.

11 | Growth of Animals

Materials

- books showing animals at different growth stages (e.g., *Bambi Grows Up, The Very Hungry Caterpillar, The Ugly Duckling, The Little Chick* (by Lisa McCue), or other books showing animal offspring and their parents)
- scissors
- glue

Activity

Read several books depicting animals and their offspring. Discuss the animals' names for offspring and adult (e.g., fawn and deer, chick and hen, caterpillar and butterfly). Ask:

- How is the adult animal the same as its offspring?
- How are they different?
- Which offspring look a lot like their parents?
- Which offspring look very different from their parents?
- How have these animals changed as they have grown?

Encourage students to describe changes in the physical characteristics of the animals they observe from offspring to adult.

Note: During discussion, explain to the students that the word *offspring* means the same as "babies" or "young." Encourage them to use the term *offspring*.

Activity Sheet

Directions to students:

Cut out the pictures and match each baby (offspring) with the animal parent (1.11.1).

Extensions

- Observe the growth of a class pet. Record observations on a regular basis. If it is possible to handle the pet, measure and graph the animal's mass weekly. Ask the students:
  - Is the animal growing?
  - What changes are taking place?
  - How big do you think the animal will get?

- Visit a hatchery to observe adult chickens and chicks. Better yet, acquire some eggs from a hatchery and hatch the chicks in an incubator right in the classroom. If you do this activity, make sure to read the book *Too Many Chickens* by Paulette Bourgeois. It is a story that focuses specifically on hatching chicks.

- In spring, collect tadpoles from a local pond or stream, then put them in an aquarium filled with pond or stream water. Feed them regular fish food. Have an elevated area in the aquarium to provide land for them once they develop. When this stage is reached, insects must be provided as food. Observe the changes in the tadpoles' stages of development and record results using diagrams and charts. Be sure to release the frogs or toads into their natural environment shortly after they are fully developed.

- Science catalogues often offer butterflies in the pupa stage. These come complete with instructions for care and background information. This activity provides students with an excellent opportunity to see a life cycle firsthand.

Growth of Animals

Date: _____   Name: _____

Growth of Animals

Offspring	Animal Parent	Offspring	Animal Parent

12 | Investigating Plants

Materials

- several live plants of various species (one for each student)
- parts of plants, such as pine cones, leaves, sunflower seeds, oranges
- knife
- plate

Activity

Begin by displaying the parts of plants. Allow the students to handle the plant parts as they observe and describe them. Ask:

- What is the same about all these things?
- What plant does each one come from?
- Are these parts alive now?

Students will likely be familiar with the term *pattern* from their activities in mathematics. Challenge them to identify the patterns on these various objects (e.g., the seeds on the pine cone, the veins on the leaf, the stripes on the sunflower seed). Have the students suggest other plant parts that have patterns (e.g., a peanut shell, a strawberry, a raspberry).

Now focus on the orange. Ask:

- Can you see the pattern on the orange?
- Is there part of the orange that you cannot see?
- What does the inside of the orange look like?
- Do you think there is a pattern inside the orange?

Carefully cut open the orange so that the cross section of the segments is visible. Discuss the pattern and challenge the students to identify other plant parts that have a pattern on the inside (e.g., watermelon, kiwi, banana, apple).

Display the live plants for students to examine. Allow the students to handle the plants as you encourage them to identify the plants and describe their characteristics. Ask:

- How are all these plants the same?
- How are they different?
- What do you know about plants?
- How do you know that they are alive?
- What do plants need in order to live?
- How do you think plants get water?
- How do you think plants get food?
- If we keep these plants in our classroom, what do we need to do to make sure that they live and grow?

Record the students' ideas about how to best care for the plants. Give each student a plant to care for. Have the students select a place where there is sunlight, and let them discuss how they will care for the plants in the weeks to come. Activity sheet A can now be completed.

Over the next several weeks, have the students care for their plants and observe changes. Times should be set aside for students to do formal observations and record changes in the plants. Activity sheet B can be used for this purpose.

Activity Sheet A

Directions to students:

Decide how you will best take care of your plant. Use pictures and words to describe your plan (1.12.1).

▶

12

Activity Sheet B

Note: There are two sheets to choose from. You may choose whichever is most appropriate for your students – an observation chart or a journal entry sheet. Alternatively, have the students select the one on which they would like to record their observations.

Directions to students:

Observe your plant and watch for any changes that happen. Use the sheet to record your observations with words and pictures (1.12.2, 1.12.3).

Extensions

- Have the students grow plants from seed and keep an observation journal of the changes that occur.

- Design investigations to see what happens to plants when they are deprived of their basic needs (e.g., place a plant in a dark cupboard, provide no water to a plant, or try to grow a plant in sand instead of soil).

- Walk around your community and observe various types of plants. Do this during different seasons so changes can be discussed. Have the students take note pads with them so they can draw or record their observations.

- Take a guided tour of a nursery, greenhouse, nature centre, or park.

- To further study the patterns on leaves, have students use the leaves for making leaf prints with paints.

Activity Centre

- Collect a large variety of pictures of plants, including coniferous and deciduous trees, shrubs, grasses, flowering plants, and fruits and vegetables. Have the students sort the pictures into groups and then challenge another student at the centre to guess their sorting rule.

- Provide the students with a variety of plant parts that have distinctive patterns and textures, such as leaves, bark, and pine cones. Have them create rubbings by colouring with a crayon over the surface of the plant parts. They can then cut out the rubbings and glue them together on paper to create texture mosaics.

Assessment Suggestion

Use informal observation to see how students are meeting the needs of their plants. Casually discuss these needs with the students to determine their understanding of the concepts and to ensure that they are providing proper care to their plants. Use the anecdotal record sheet on page 15 to record results.

Date: _____   Name: _____

Caring for My Plant

My plant needs_____

_____

_____

_____

_____

Date: _____   Name: _____

Observing My Plant

Date	Observations

Date: _____

Observations:

| | | | | |

My Plant
Journal
by
_____

13 Maintaining a Healthy Environment

Materials

- chart paper, felt pens
- mural paper
- large pictures of the environment (included) (1.13.1-1.13.8)
- scissors, glue

Note: You may wish to provide the students with valuable background information on environmental issues prior to beginning the activity. Read appropriate books about pollution, extinct animals, forest fires, and other related issues. You may also wish to show videos or have guest speakers in to discuss environmental issues. Ensure that such books and other resources are appropriate for grade-one students.

Activity

Begin by displaying the eight pictures of healthy and harmful environments (1.13.1-1.13.8). Have students closely examine each picture, then discuss them. Explain that some of the pictures show good things that help keep humans and other living things healthy.

Explain that the rest of the pictures show bad activities that are harmful to humans and other living things. Ask:

- Which pictures show healthy environments?
- Why are they healthy?
- Which pictures show harmful environments?
- Why are they harmful?

Have the students sort the pictures according to healthy environments and harmful environments.

Review the needs that humans have in order to live. Focus on health issues by asking:

- What do you need to do to stay healthy?
- How do humans get sick?
- How can you take care of yourself so you do not get sick so often?

Throughout the discussion ensure that you talk about eating properly, washing hands, exercising, getting enough sleep, and dressing properly for weather conditions. You should also discuss environmental issues such as clean air and water.

Divide chart paper into three columns. At the top of the first column, print the title Taking Care of Ourselves. Record the students' ideas under this heading.

Now extend the discussion to include keeping other living things healthy. Ask the students:

- How are animals important to us?
- How do humans sometimes hurt animals?
- How can we take care of animals to make sure that they are alive and healthy like us?

At the top of the second column of the chart, print the title Taking Care of Animals, and record the students' ideas. This list might include issues such as keeping lakes clean so that animals have water, limiting hunting, reducing pollution, and maintaining natural habitats for animals.

Now focus on plants by asking the following questions:

- How are plants important to us?
- How do humans sometimes hurt plants?
- How can we take care of plants to make sure that they are alive and healthy like us?

At the top of the third column of the chart, print the title Taking Care of Plants, and record students' ideas. This list might include preventing forest fires, not cutting down all the trees, and keeping the air and water clean.

Divide the class into three groups and have each group make a mural depicting ideas from one of the three columns from the chart. Display the murals for all to see.

▶

13

Activity Sheet

Directions to students:

Cut out the pictures and sort them into two groups, one showing harmful environments and the other healthy environments. Glue the pictures under the correct headings (1.13.9).

Note: The pictures that show a healthy environment are: boy washing hands with soap and water, recycling box, birds at a bird feeder, girl putting garbage in garbage container. Pictures that show ways to harm the environment are: boy sneezing without covering nose/mouth, broken glass in playground, bulldozer uprooting trees, building emitting pollution into the air.

Extensions

- Read the book *For the Love of Our Earth* by P.K. Hallinan. It is an excellent resource for environmental issues and is very appropriate for this grade level.

- Create posters using the theme Keep Our Environment Healthy. Display the posters around the school for all to see.

- Make a class book using the same theme, Keep Our Environment Healthy. Have the students illustrate ways humans can protect their environments. Collect the pictures, bind and cover the book, and place it in the class library.

- Visit a local wildlife sanctuary to learn firsthand how animals and plants are protected.

- Initiate a class project for collecting recyclable items such as pop bottles, newspaper, and tin cans. Encourage the students to recycle at home as well.

- In the classroom, collect paper instead of throwing it in the garbage. It can be used as scrap paper for drawing on or for art projects. Most schools have a paper recycling program, so the students can also collect the paper for this purpose.

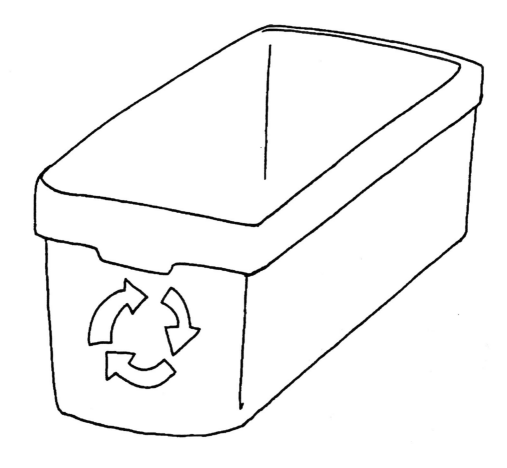

The Environment – Healthy or Harmful?

The Environment – Healthy or Harmful?

Healthy Environments	Harmful Environments

References for Teachers

Althouse, Rosemary, *Investigating Science with Young Children*. New York: Teachers' College Press, 1988.

Ballard, Carol. *How Do Our Eyes See?* Austin: Raintree Steck-Vaughn, 1998.

_____. *How Do Our Ears Hear?* Austin: Raintree Steck-Vaughn, 1998.

Bosak, Susan. *Science Is... .* Richmond Hill, ON: Scholastic, 1991.

Butzow, Carol, and John Butzow. *Science Through Children's Literature*. Englewood: Teacher Ideas Press, 1989.

Englehart, Deirdre. *The Five Senses.* Grand Rapids: Instructional Fair Publishing Group, 1999.

Enns, Elfreda. *Our Five Senses.* B.C. Teachers' Federation Lesson Aids Service, n.d.

Flagg, Ann. *Our Bodies.* Huntington Beach: Teacher Created Materials, 1995.

Liddelow, Lorelei. *Cook With Me.* Winnipeg, MB: Peguis Publishers, 1990.

McCracken, Robert, and Marlene McCracken. *Animals*. Themes. Winnipeg, MB: Peguis Publishers, 1985.

_____. *The Sea & Other Water*. Themes. Winnipeg, MB: Peguis Publishers, 1985.

_____. *Myself*. Themes. Winnipeg, MB: Peguis Publishers, 1985.

Suzuki, David. *Looking at Plants.* New York: Wiley, 1991.

Trostle, Susan Louise. *Integrated Learning Activities for Young Students.* Needham Heights: Allyn and Bacon, 1990.

Unit 2

Characteristics of Objects and Properties of Materials

Books for Children

Blegvad, Erik. *The Three Little Pigs*. New York: Atheneum, 1980.

Brett, Jan. *Goldilocks and the Three Bears*. New York: Dodd, Mead, 1987.

Graves, Kimberlee. *I Can't Sleep*. Learn to Read Science Series, Level 2. Cypress, CA: Creative Teaching Press, 1994.

_____. *What's In My Pocket?* Learn to Read Science Series, Level 2. Cypress, CA: Creative Teaching Press, 1994.

_____. *Mom Can Fix Anything*. Learn to Read Science Series, Level 2. Cypress, CA: Creative Teaching Press, 1994.

Medearis, Angela Shelf. *We Play on a Rainy Day*. New York: Cartwheel Books, 1996.

Munsch, Robert. *Mud Puddle*. Toronto: Annick Press, 1982.

Simon, Norma. *Wet World*. Cambridge, MA: Candlewick Press, 1995.

Thrall Cicciarelli, Joellyn. *Apron Annie in the Garden*. Learn to Read Science Series, Level 3. Cypress, CA: Creative Teaching Press, 1997.

Williams, Rozanne Lanczak. *Reduce, Reuse, Recycle*. Learn to Read Science Series, Level 2. Cypress, CA: Creative Teaching Press, 1994.

_____. *Buttons, Buttons*. Learn to Read Science Series, Level 2. Cypress, CA: Creative Teaching Press, 1994.

Creative Teaching Press books are available from Peguis Publishers, Winnipeg

Web Sites

- **http://www.exploratorium.edu/IFI/ activities/iceballoons/figure2.html**

  Exploratorium Institute for Inquiry outlines various ways to encourage questions in the classroom – specifically for science investigation.

- **http://www.owu.edu/~mggrote/pp/**

  Project Primary is a collaborative effort between elementary school teachers and professors from six different departments at the University of Illinois. Click on "Chemistry" to find information and activities on polymers, kitchen chemistry, and nitro ice-cream.

- **http://www.sci.mus.mn.us/sln/tf/**

  Science Museum of Minnesota's Thinking Fountain: click on "Theme Clusters" and search for "Loose Parts," where you can find activity ideas for your collection of recycled materials. "Helpful Tidbits," under the same heading, offers tips for teachers for understanding concepts such as density, mass, and volume.

- **http://www.epa.gov./recyclecity/**

  From the United States Environmental Protection Agency: an interactive site for teachers and students; click anywhere on "Recycle City" to find ways to reduce, reuse, and recycle. With links for teachers and students.

- **http://www.flash.net/~spartech/Reeko Science/ReekoIndex.htm**

  Reeko's Mad Scientist Lab from Discovery Online: fun and informative science experiments for teachers and students on mass, density, and volume. With extensive links.

- **http://www.epsea.org/adobe.html**

  Adobe Home Construction for the El Paso Solar Energy Association: information on the construction of adobe houses, the materials used, and the structures' history. With links to straw bale homes, resources on energy efficiency, and solar heating.

Introduction

In this unit, students are introduced to concepts about materials through exploration of objects in their immediate surroundings. Students will use their senses to identify various objects and materials. In doing this, they will learn to make a clear distinction between objects and materials: they will learn that objects are made from materials and that materials have specific properties. They will also learn to describe these properties clearly and precisely. By making objects out of various materials, they will begin to understand that there is a connection between the properties of materials and the specific purposes for which the materials are used.

During all activities, it is important to teach students about proper scientific testing procedures. For example, when smelling objects, students should not inhale the fumes directly, but should wave a hand over the object toward the nostrils. In addition, students should not taste objects during science experiments unless they know that it is safe and they have been given adult permission.

General Information for Teachers

Anything that takes up space and has weight is called *matter*. Colour, odour, taste, weight, and hardness are some of the physical properties of matter.

Science Vocabulary

Throughout this unit, teachers should use, and encourage students to use, vocabulary such as: *object, material, wood, metal, plastic, cloth, waterproof, absorbent, rigid, flexible, strong, weak,* and *recycle.*

Materials Required for the Unit

Classroom: plastic overhead transparency sheet, chalk, pens, crayons, scissors, materials in the classroom, index cards, markers, chart paper, water, paint, stapler, masking tape, paper clips, Scotch tape, string, wool, glue, elastic bands, pencils, paper, labels, poster paper

Books, Pictures, and Illustrations: a copy of *Goldilocks and the Three Bears*, *We Play on a Rainy Day* (a book by Angela Shelf Medearis), *Wet World* (a book by Norma Simon), *Mud Puddle* (a book by Robert Munsch)

Household: running shoes, cloth, paper towels, household items made from different materials (e.g., wood, metal, plastic, rubber, paper, cloth), bowls, apple, potato, several kinds of knives (e.g., paring, peeler, plastic, butter), plastic wrap, rain coats, rubber boots, umbrellas, eyedroppers, cloth samples (cotton, rayon, linen, waterproof nylon and plastic), egg cartons, Styrofoam trays, Styrofoam plates, aluminum foil, Styrofoam cups, paper plates, napkins, toothpicks, plastic and paper bags, boxes, plastic pop bottles, plastic milk jugs, newspaper, wrapping paper, margarine tubs and other plastic food containers, cardboard tubes, tinfoil, pizza recipe, milk cartons

Other: Hula-Hoops, sandpaper, foam, soft fabric, cloth-covered chair, metal chair, wooden chair, stool, straw, sticks or Popsicle sticks, Lego, a brick, wooden block, toy drum (or other musical instrument), chocolate (e.g., Smarties), banana, box or blanket, pizza dough, shredded cheese, pizza cutter, pizza sauce, tomato sauce, oven, small finishing nails, uncooked rice, sparkles

1 The Five Senses

Materials

- wooden block
- toy drum or other musical instrument
- running shoes
- chocolate (e.g., Smarties, one for each student)
- peeled ripe banana (place the banana and the peel in a container with small holes; students will be able to smell the banana, but not see it)
- box or blanket

Note: Be aware of student allergies when conducting taste tests. Be especially cautious in cases of peanut allergies – many brands of chocolate contain traces of peanuts.

Activity

Begin by having all the objects hidden from view in a box or underneath a blanket.

Show the students the wooden block. Have the students look at it and identify what it is. Ask:

- How did you know what this object was?
- What part of your body do you see with?

Now have the students close their eyes. Tap on the drum or play another musical instrument. Challenge the students to identify what the object is. Ask:

- How did you know what the object was even though your eyes were closed?
- What part of your body do you hear with?

Have the students close their eyes again and open their mouths. Place the chocolate on their tongues (to avoid spreading germs, make sure you wear gloves or have just washed your hands). Have the students identify what is on their tongue. Ask:

- How did you know that it was a Smarties?
- What part of your body do you taste with?

Once again, have the students close their eyes. Pass around the running shoes. Encourage students to identify the item, but not to call out the answer until everyone has held it. Have the students identify what was in their hands. Ask:

- How did you know it was a running shoe?
- What part of your body do you smell with?

Now pass around the container with the banana in it. Have the students smell the odour by waving their hand over the holes toward their face. Again, encourage students to identify the object, but not aloud until everyone has had a chance to smell it. Ask:

- How did you know it was a banana?
- What part of your body do you smell with?

Note: The activity sheet is to be used during the activity.

Activity Sheet

Directions to students:

Print the name of, or draw, each object in the first column. In the second column, draw diagrams to show which sense you used to tell what each object was (2.1.1).

Activity Centre

Have a collection of objects at the centre that can be identified using the senses:

- touch: uncooked macaroni, feathers, buttons, and paper clips
- hearing: several sounds recorded on a tape recorder
- smell: lemon or orange peels, chocolate drink powder, and fresh ground coffee
- taste: crackers, raisins, cereal, and popcorn

1

Note: Be aware of student allergies during these activities, and select items accordingly.

Students can work in pairs. One student can be blindfolded, then try to identify objects with his or her senses, while the other guides the activity. Have the students record their findings at the centre on the activity centre sheet (2.1.2).

Extensions

- As a class, play a game such as I Spy or Twenty Questions. Have the students use their senses to identify objects.

- During art activities such as finger painting, have students describe the feeling of the paint and the paper they are using. Try other media to paint on including cloth, canvas, plastic, and wood. Ask students to describe how the various materials feel.

Assessment Suggestion

Observe students as they work cooperatively at the activity centre. Use the anecdotal record sheet on page 15 to record results of the students' understanding of the concepts, and of their ability to work together.

Using My Senses

Object	Sense

Date: _____ **Name:** _____

What Is the Object?

Sense Used	Object

2 Observing and Describing Materials

Science Background Information for Teachers

It is important that students use appropriate language to describe their observations about materials. Encourage them to use descriptive words and phrases such as *rough, smooth, shiny, dull, spongy,* and *soft.*

Materials

- poster paper
- glue
- sandpaper, foam, soft fabric (such as velvet), tinfoil, and overhead transparency sheets (cut pieces of each material in 10 cm squares) (Have enough materials so that each student has three samples of each material.)
- chalk
- markers
- crayons

Activity

Provide all students with the squares of sandpaper, foam, fabric, tinfoil, and transparency sheets. As a group, discuss the characteristics of each material, focusing on how it feels and looks. Have the students glue the three samples of each material onto a sheet of poster paper, and label the poster as in the following example:

	sandpaper	foam	fabric	tinfoil	plastic
chalk					
marker					
crayon					

Provide students with chalk, markers, and crayons. Explain that their task is to use these writing instruments to draw on their squares of material. Ask:

- How well do you think chalk will work for drawing on sandpaper?
- Will chalk work well on foam (fabric, tinfoil, plastic)?
- How well do you think markers will work on each material?
- Will crayons be good for drawing on these materials?

Challenge students to use the chalk, markers, and crayons to draw pictures, print their names, or write messages on the material samples. As they work, encourage them to discuss their findings and record their observations on their activity sheet.

Following this activity, have the students share the results and display their posters for the class. Discuss the characteristics of each material and how well the writing devices worked for drawing and printing (e.g., chalk marks well on a rough surface such as sandpaper, but does not mark well on a smooth surface such as tinfoil or plastic).

Activity Sheet

Directions to students:

Print the name of each material on the chart. Use a checkmark if the chalk, marker, and crayon worked well for drawing and printing on that material. Use an X if the chalk, marker, and crayon did not work well for drawing and printing on that material (2.2.1).

Extension

Give students plenty of opportunities to manipulate, observe, and describe objects. This type of activity also develops important

▶

2

language skills. Have the students select an object in the classroom, make up a riddle describing the objects, and share the riddle with a classmate to see if the classmate can guess the object. (For example: My object is shiny and smooth, it is silver, and it is sharp. What is my object?)

Activity Centre

■ Gather samples of materials such as wax paper, tinfoil, rug samples, cloth, plastic grocery bags, plastic transparency sheets, sandpaper, paper towels, and so on. Label each object so students can refer to these words when completing the activity centre sheet. Have the students examine the materials and sort them by texture (e.g., fuzzy, smooth, soft, scratchy). They will then complete the activity centre sheet (2.2.2) by cutting out a small piece of the

material and gluing or taping it onto the chart. Students can then describe each piece of material. (You can also display a list of words on chart paper at the centre for students to refer to when writing.)

■ Provide several containers of objects (e.g., buttons, bottle caps, small toys, geo-shapes, fabric samples) and paper plates. Encourage students to classify the objects according to their own rules. Also provide index cards so the students can record their sorting rules. To further extend the activity, have additional index cards with various sorting rules on them, especially those rules that students have not regularly used (e.g., pointed and not pointed, curved and straight edges). These index cards can have picture clues on them so students do not have to rely solely on reading.

Date: _____ Name: _____

Drawing on Materials
☑ or ☒

Material	Chalk	Marker	Crayon

Date: _____   Name: _____

Materials

Sample of the Object or Material	Describing Words

3 | Classifying Materials

Materials

- several household items made from different materials (wood, metal, plastic, rubber, paper, cloth)
- materials in the classroom
- Hula-Hoops
- pencils
- index cards
- markers

Activities

- Have the students form a circle. Place the household items on the floor in the middle of the circle. Pass the items around so that students have an opportunity to examine them and describe them.

Have the students sort the objects according to which material they are made from, then place the objects inside the Hula-Hoops according to material. The students will notice that some objects are made of more than one material and must be left out of the sorted groups (or additional groups can be made for these objects). You may also want to introduce intersecting Venn diagrams during this activity. For example:

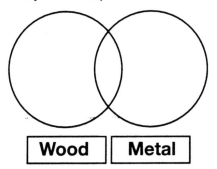

- Have a scavenger hunt and challenge the students to find objects made of specific materials from around the classroom.

The students will use the activity sheet to record their findings.

Activity Sheet

Directions to students:

During the scavenger hunt, draw a picture of the objects you find that are made of these materials (2.3.1).

Extension

Have the students take an activity sheet home and complete it with drawings of objects found around their home.

Activity Centre

Provide construction paper, scissors, and glue, along with a variety of objects made from different materials such as:

- wood (e.g., toothpicks, bark, wood shavings)
- plastic (e.g., cutlery, plastic wrap, bread tags, bingo chips)
- metal (e.g., paper clips, staples, aluminum foil, nails, screws, bolts)
- cloth (e.g., yarn, wool, thread, various types of fabric)
- paper (e.g., plates, napkins, wrapping paper, stickers)

Have students create collages made of only one material. Have them title their pictures with the name of the material used.

Set up several activities at the centre so students have more opportunities to observe and describe materials. Have them do the following:

- Test wood pieces for softness by scratching the pieces with a nail or by hitting the wood with a hammer.

3

- Rate metals for shininess by ordering them from shiniest to dullest.
- Test rock samples for hardness with a nail scratch test.
- Compare the thickness of various types of cloth.
- Compare fabric samples for smoothness.

Assessment Suggestion

Observe students as they describe and classify materials during the class activity and scavenger hunt, and when they are at the activity centre. Use the anecdotal record sheet on page 15 to record results.

Name: _____

Date: _____

Objects and Materials

Paper				
Wood				
Metal				
Plastic				
Cloth				
Glass				

4 | Properties of Materials

Materials

- 4 bowls, labelled A, B, C, and D
- apple
- potato
- several different types of knives (e.g., paring, peeler, plastic, and butter knife) (for teacher use only)
- plastic wrap
- chart paper
- markers

Activity

Display the apple and potato for students to observe and examine. Ask them to describe the inside and outside of each.

Note: During the next component of the activity, make sure to review safety issues when using knives, and stress that students should not use sharp utensils.

Display the various knives, have the students describe each, and predict which one would be best for peeling the apple and the potato. Test their predictions by trying to peel the potato with each of the four knives. Ask:

- Which knife works best? Why?

Complete the peeling of the apple and the potato.

Now ask the students to predict which knife would be best for slicing the apple and the potato. Test their predictions by trying to slice the apple with each of the four knives. Ask:

- Which knife works best? Why?

Complete the slicing of the apple and the potato.

Have the students observe and examine the slices. Ask:

- What do you think will happen if you leave these on the table for a long time?
- How do you know that will happen?
- Why do you think the apple and potato will turn brown?
- Can you think of any way that you could stop the apple and the potato from turning brown? What could you use?

Place half the potato pieces in bowl A, uncovered. Wrap the other half of the potato slices in plastic wrap and place them in bowl B. Do the same with the apple pieces, using bowls C and D.

Divide chart paper into two columns. At the top of the first column, print the heading Predictions (explain to the students that the term *prediction* means making a guess about what will happen). Have the students predict what will happen to the apple and potato slices if the slices are left in the bowls for one hour. Record their predictions on the chart.

Have the students complete the first column of the activity sheet by drawing pictures to show what the slices in each of the four bowls look like before the hour has passed.

Continue with other classroom activities until the hour has passed, then have the students examine the apples and potatoes again. At the top of the second column of the chart, print the heading Results and explain to the students that the term *results* means what really happened in an experiment. Record their observations.

Have the students complete the second column of the activity sheet by drawing pictures to show what the apple and potato slices in each of the four bowls look like after the hour has passed.

4

Have the students infer why the apples and potatoes wrapped in plastic wrap did not turn brown. Discuss the characteristics of the plastic wrap that make it useful in preventing the apple and potato slices from turning brown quickly. Also discuss other ways that plastic wrap is used in the home (e.g., wrapping up sandwiches and other foods for lunch).

Note: Students do not need to provide a complex scientific explanation during this activity. Simply, when apples and potatoes are covered with plastic wrap, they will not turn brown quickly because the air cannot get at them.

Activity Sheet

Directions to students:

Use drawings to show what each of the four bowls of food looks like before the experiment. Then draw the results of the experiment (2.4.1).

Extensions

- Conduct investigations to see what other materials would stop the apple and the potato slices from turning brown (e.g., cheesecloth, plastic bags, paper bags, aluminum foil, wax paper). Have students compare and order materials according to how well thay work and seal the apple and potatoe slices from air contact.

- Play What If, focusing on properties of materials. Ask the students to imagine what would happen in the following situations:

  - What if a ball were made of rock?
  - What if your jacket were made of wood?
  - What if your kitchen pots were made of paper?
  - What if your mittens were made of aluminum foil?

  As each question is answered by students, discuss the materials that the items are really made from and the properties of those materials (e.g., a ball is made from rubber because rubber bounces well). Also encourage the students to create their own "What If" questions.

- Have students examine and manipulate carpentry tools. Focus on the function of each tool and the materials each is made of. Include hammer, screwdriver, sandpaper, nail, tape measure, pliers, and so on.

Experimenting With Food

Before	After
A	A
B	B
C	C
D	D

5 | Comparing Objects

Materials

- copy of the book *Goldilocks and the Three Bears*
- cloth-covered chair
- metal chair
- wooden chair
- stool

Activity

Read the book *Goldilocks and the Three Bears*. Discuss the story and focus on the chairs that Goldilocks sat on. Ask:

- What was wrong with Father Bear's chair?
- What was wrong with Mother Bear's chair?
- Why do you think Baby Bear's chair was just right?
- What do you think the chairs in the story are made of?

Now present the four chairs to the students. Ask:

- What are chairs used for?
- How are all these chairs the same?
- How are they different?
- What is each chair made from?
- Where does wood come from?
- Which do you think is the strongest chair? Why?

Have the students examine each chair and record on the activity sheet the similarities and differences, including materials used and type of construction.

Activity Sheet

Directions to students:

Observe the chairs. Draw a picture of each chair in the first column. For each chair, tell how many legs it has. Then use a checkmark to show if the chair has a back on it. Now use a checkmark to show if the chair is made of wood, metal, or cloth (2.5.1).

Extensions

- Have the students order the chairs according to strength or comfort.
- Have the students complete the extension activity sheet (2.5.2) at home by drawing pictures of the different doors in their homes and observing the characteristics and materials used in each door.
- Focus again on the story *Goldilocks and the Three Bears*, and ask the students about the beds:
  - What was wrong with Father Bear's bed?
  - What was wrong with Mother Bear's bed?
  - Why do you think Baby Bear's bed was just right?
  - What do you think the beds in the story are made of?

Explain to the students that they are going to have a chance to make a bed for a stuffed animal. Ask them to bring their favourite stuffed animal to school. Tell them they are going to design a bed for their animal. When all the students have brought in a stuffed animal, have a group discussion to examine and describe the animals. Focus on things that the students will need to keep in mind when making a bed, such as the size of the stuffed animal and how comfortable the bed is for it.

Have the students collect items from home (or supply as many as you can in the classroom), such as boxes, pillow stuffing, cloth, and packing tape. Students can work on constructing the beds at home, during class time, or at the activity centre. Plan a celebration of learning when students can present their finished products to the class.

Comparing Chairs

Chair	Number of Legs	Back	Wood	Metal	Cloth

Date: _____ Name: _____

Comparing Doors

	Number of Windows	Handle	Wood	Metal	Screen
Front Outside Door					
Bedroom Door					
Back Door					
Another Door					

6 | Changes in Materials

Science Background Information for Teachers

There are two main types of changes that can occur in materials: physical and chemical. Cutting paper is an example of physical change; only the size and shape have changed. Chemical change makes a new kind of matter. When a log is burned, ashes are left. The molecules of the wood change.

Materials

- pizza dough
- shredded cheese
- pizza/tomato sauce
- oven
- pizza recipe
- pizza cutter
- paper plates, napkins

Activity

Note: Be sure to review with students the importance of washing their hands before handling food.

Have students examine the ingredients before putting the pizza together. Ask:

- How does the pizza dough look before it is cooked?
- How does the sauce look?
- How does the cheese look?
- How do they each smell?
- How do they each feel?

After the students make the pizza, have them predict what will happen while the pizza is baking. Ask:

- What will the heat in the oven do to the crust?
- What will the heat do to the sauce?
- What will happen to the cheese as it is heated?

While the pizza is baking, have the students complete the activity sheet so that they have a pizza recipe to take home.

When the pizza has been baked and cooled, give the students an opportunity to enjoy a slice. As they are eating, have them compare their prediction about what they thought would happen to the pizza while it was baking to how the pizza actually did change with heating.

Activity Sheet

Directions to students:

Use the sheet to make a recipe for your pizza. Draw pictures and use words to tell about the ingredients (what you used) and the directions (what you did) (2.6.1).

Extension

Give students a piece of Play-Doh and have them roll it into a ball. Observe and discuss the form, shape, and colour of the Play-Doh. Now ask the students to change the Play-Doh in some way (e.g., the shape). Ask them if they can change it back to the way it was before. Now give students a piece of Play-Doh of a different colour. Challenge them to find a way to use both pieces of Play-Doh and change them so that they cannot be changed back to the way they were before (e.g., mix the colours).

Brainstorm ways that Play-Doh can be changed: by heating or cooling it, adding water or salt, mixing the Play-Doh with beads, and so on. Have the students predict which changes can be reversed. Record their predictions and then plan investigations to determine the results of these changes.

Have the students try to change other materials such as paper, aluminum foil, cloth, crayons, and sticks.

6

Activity Centre

Provide several small pieces of plywood, paint, and paintbrushes. Have the students examine and describe their piece of wood, observing the smell and texture of it. Now have the students paint the entire piece of wood, adding decorative pictures. After the paint has dried, have them again observe the smell of the wood and the smoothness/roughness of the wood. While students are working at the centre, ask them to describe how the wood has changed and to infer the reason for this change.

Pizza Recipe: What We Used

_____

_____

_____

_____

_____

_____

Pizza Directions: What We Did

_____

_____

_____

_____

_____

7 | Materials and Sound

Science Background Information for Teachers

Rain sticks come from South America. The people there cut sections of cactus, pull off the sharp edges of the thorns, and push the stems back through the soft outer layer of the cactus. The cactus is placed in the sun to dry, and then filled with pebbles. Both ends are then covered with wood. If you turn the rain stick up and down, the pebbles will run through the centre of the stick hitting the thorn stems, causing them to vibrate. This vibration is what creates sound.

Materials

Each student will need:

- cardboard tube (from paper towel or gift wrap)
- 30 round toothpicks (break them in half if they are longer than the diameter of your tubes)
- small finishing nail
- paper
- scissors
- masking tape
- uncooked rice

Activity

Making Rain Sticks

Note: This activity will require some help from parent volunteers or older students. If you have a buddy reading program with a class from a higher grade level, you may want to ask for their assistance on this special project. Both groups of students will enjoy the experience.

Have the adult/older student use the nail to poke about thirty holes throughout the surface of the tube. Place the toothpicks through the holes, leaving a little bit sticking out. Wrap the entire tube with masking tape (packing tape works well also) so the toothpicks do not fall

out. Cut out two circles of paper and tape one in place over the bottom hole of the tube. Pour a handful of uncooked rice into the tube. Cover the top with the second circle and tape it into place. Now turn the tube upside down and listen. When the sound stops, turn the tube over again.

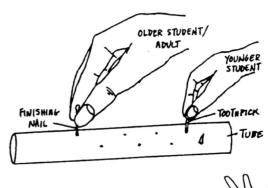

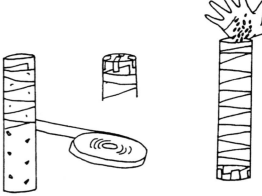

Note: Before sealing the top, cover the top of the tube with your hand and turn the tube over. Add or remove rice until you get the sound you like, then seal it with the cover. Dry soup mix, dry beans, and sand will produce different sounds.

When students have all had an opportunity to try out their rain sticks, ask:

- How was the sound made?
- Were you able to change the sound? How?
- What else could you make rain sticks out of to get different sounds?

▶

7

Extensions

- Challenge the students to make rain sticks out of other materials to investigate how the materials change the sounds produced. Find out why the people in South America make rain sticks. This is a good opportunity to invite a guest speaker to the class, especially if that person has an authentic rain stick.

- Make tin can drums. Collect recycled, cleaned, food cans, two for each student (round-bottom cans will not work). Remove the top and bottom from half the cans and just the top from the rest of the cans (it is best to tape the sharp edges of all cans to avoid injuries).

Note: Students may need individual assistance in constructing their drums, but the follow-up activity can be done as a group.

Using packing or duct tape, have the students tape the two cans, open ends together.

Note: Two cans are used to make the drum beacuse a better echo chamber is created. You can also investigate what happens if more cans are connected to make longer drums.

Now it is time to experiment with sound. Provide the following instructions to students:

- With a pencil, tap on the side of the drum.
- Place a towel on the floor, put the drum on top of it, and tap the drum again.
- Hit the drum slowly, quickly, and lightly.
- Run the pencil up and down the side of the drum like a washboard.
- Sing or talk into the open end of your drum.
- Tap your hand on the closed end of the drum.

Discuss the results as students attempt each task. Ask:

- What types of sounds did you produce?
- How did the sounds change?

Note: Different-size cans will produce a different pitch.

8 | Testing Materials

Materials

- story dealing with rain, such as *We Play on a Rainy Day* by Angela Shelf Medearis, *Wet World* by Norma Simon, or *Mud Puddle* by Robert Munsch
- sample raincoats, rubber boots, and umbrellas
- eyedroppers
- water
- paper towels
- fine types of materials that vary in their ability to absorb water (e.g., cotton, rayon, linen, and waterproof nylon and plastic) (Label the samples.)
- chart paper
- markers

Activity

Begin by reading a story about rain to the class. Discuss the book and ask:

- What do you do to stay dry when it rains?
- How do you dress when it rains?
- Why does the water not go through your raincoat?

Have the students examine the sample raincoats, rubber boots, and umbrella (to avoid injuries, have students examine the umbrellas without opening them). Ask:

- What are the materials like?
- How are they different from the clothes you are wearing now?

Divide the class into cooperative groups, then have the groups examine and describe the various kinds of cloth samples. Have the groups predict and sort the cloth samples according to which would keep out water and which would let the water through.

Divide chart paper into two columns. Title the first column Predictions and record the

students' predictions on chart paper. Introduce the word *absorb*. Explain that when a material absorbs water it means that the material soaks up the water.

Now have the students test their predictions. First, let them practice using the eyedroppers. They can do this by counting out five drops of water on a paper towel. Now have them place each cloth sample on a paper towel, then squeeze a total of five drops of water on each sample. Students can then lift the samples to observe through which ones water leaked onto the paper towel. Record the results on the chart. Ask:

- Which samples absorb the most water?
- Which samples absorb only a bit of water?
- Which samples do not absorb the water?
- If you had to make your own raincoat or umbrella, which materials would you use?

Now have the students repeat the water-drop experiment using ten drops of water on the cloth samples. Predict and test to see if the results from the previous experiment change. Record results.

Activity Sheet

Directions to students:

Print the names of the different materials in the first column. Use a checkmark to show if the water soaked through each material and an X to show if it did not (2.8.1).

Extensions

- Have students try the water-drop experiment at home to test how other materials absorb water. Make sure they check with their parents first. They could test objects such as tumblers, plates, socks, gloves, or jeans. Have them report back to the class.

8

- Observe, compare, and test the insulating properties of several materials. Take five small Ziplock bags and put four ice cubes into each bag. Seal the bags. Now fill four large Ziplock bags with materials such as feathers, Styrofoam chips, cotton batting, and aluminum foil (one type of material per bag). Place a bag of ice inside each of the larger bags. Try to have about 3 cm of insulating material around the ice. Another interesting insulator to test is air: place the fifth bag of ice into a large bag filled with air only. Have students predict how the outer bags will feel (cold, warm, hot). After a few minutes, have the students feel the outer bags to determine how well the materials are insulating. Leave the bags for several hours. Check them regularly to see how long it takes the ice cubes to melt.

Note: Introduce the science concept regarding controlling variables. For students, this means making the test as fair as possible. Therefore, it is important to ensure that, at least approximately, there are equal amounts of materials placed inside each of the large bags.

Activity Centre

Organize a centre where students can test the insulating properties of materials. Provide a variety of materials such as shoeboxes, plastic containers, metal containers, fabric samples, cotton batting, wood shavings, shredded paper, and so on. Challenge the students to design a container that will keep an ice cube frozen for the longest time possible.

Over a period of days, have each student in the class design and construct an ice cube keeper. When all of the ice cube keepers have been made, have the students examine, compare, and contrast them. Also have the students predict which ice cube keeper will keep the ice cube frozen for the longest period of time. Record their predictions and set a date for the test. Place an ice cube in each container at the beginning of the day and check on the ice cube every hour to determine the results. Record these results and have the students infer why some ice cube keepers worked better than others (e.g., Why do you think the ice cube melted in this ice cube keeper so quickly? Why do you think this other ice cube keeper kept the ice cube frozen for four hours?) Compare the ice cube keepers to items used to keep food cool, such as thermoses, coolers, and thermal lunch bags.

Assessment Suggestions

- Observe the students as they conduct the water-drop tests. Check for: (1) accuracy, (2) counting, (3) observational skills, (4) students' ability to describe results, and (5) students' ability to determine the best material for a raincoat or an umbrella for repelling water. List these criteria on the rubric on page 19 and record results.

- While students are working at the activity centre building their ice cube keepers, observe their ability to plan and construct the item, as well as their final design. Use the anecdotal record sheet on page 15 to record results.

Date: _____ Name: _____

Using Water to Test Materials

Material	5 Drops	10 Drops

9 | The Purpose and Function of Materials

Materials

- several versions of the story *The Three Little Pigs*
- straw
- sticks or Popsicle sticks
- Lego (to be used as bricks)
- brick
- collection of materials that can be used to join or fasten other materials (stapler, masking tape, Scotch tape, paper clips, string, wool, glue, elastic bands, and so on)

Activity

Read various versions of *The Three Little Pigs* to the students. Discuss the story, focusing on the materials that the houses were made of.

Pass around the straw and the brick for students to examine and manipulate. Encourage students to describe the properties of each material. Ask:

- What does the straw/brick look like?
- What does the straw/brick feel like?
- What do humans use straw/brick for?
- Where does straw/brick come from?

Now focus on the houses built by the three pigs. Ask:

- What objects did the three pigs build?
- What did the first pig build his house out of?
- What did the second pig build his house out of?
- What did the third pig build his house out of?
- Which pig chose the best material with which to build his house? Why is it the best material?

- Why could the wolf not blow down the house made of bricks?
- Why do you think the house made from bricks was so strong?
- Why were the houses made of straw and sticks not as strong?

Divide the class into working groups. Have each group design and construct houses made of straw, sticks, and Lego bricks.

As they are building, encourage students to discuss the properties and characteristics of the materials they are using. Also encourage them to find ways to fasten the materials together so that the houses are as sturdy as possible.

Provide an opportunity for the groups to present their houses to the class. Invite them to discuss the materials they used, the properties of those materials, the fasteners they used, and the sturdiness of their houses.

Activity Sheet

Directions to students:

Draw a diagram of each house that your group built. Put the houses in order from weakest to strongest (2.9.1).

Extensions

- Display pictures of different kinds of houses, then discuss the materials used to build each house as well as the characteristics of each. Take the students on a walk around the neighbourhood to look at houses and identify the materials used in the construction of each house.

- Challenge students to build a playhouse for the classroom, using a large refrigerator

▶

9

box. As a class, brainstorm the features that the house would need, such as doors, windows, stairs, and curtains. Students may also want to include features such as a cozy reading corner, window boxes, and carpeting. Have students design the playhouse, collect the materials to construct it, then build it at the activity centre. They may want to use the playhouse as a permanent reading centre or as a play centre.

Date: _____

Name: _____

Building Houses

Weakest House

House made of _____

Strongest House

House made of _____

House made of _____

10 | Designing and Constructing Objects

Materials

Note: This is an opportunity for students to learn about recycling by collecting items from home. Send a note home to parents explaining that their children are going to be doing a special project building items made from cleaned, recycled materials.

Suggest an extensive list of materials to students and parents. These might include:

- egg cartons
- Styrofoam trays, plates, and cups
- plastic and paper bags
- boxes
- plastic pop bottles and milk jugs
- newspaper and wrapping paper
- cardboard milk cartons
- margarine tubs and other plastic food containers
- cardboard tubes
- other items, such as tape, glue, fasteners, paint, aluminum foil, and sparkles (for decoration), that students determine after planning their project

You will also need the following items:

- several boxes for sorting items
- chart paper
- markers

Activity

As items are brought to school, have the students sort them into the boxes. When you have the quantity and variety needed, display the items for students to examine. Ask:

- Where did these objects come from?
- Why did you not just throw them in the garbage?
- Why is it important to reuse and recycle?
- How does this help the environment?

Now explain to the students that they are going to begin a special project making a new object out of the materials displayed. Have them brainstorm things that they can make. The list may include the following:

- pencil holder
- crayon box
- desk organizer
- boat
- bird feeder
- birdhouse
- car
- toy animal or doll
- rocket
- spaceship
- robot
- characters from books they have read

Note: You may also refer to children's craft books for further ideas on items to make.

Give students plenty of time to discuss how, using these materials, various items might be made (e.g., a boat could be made from a milk carton, with a straw and a piece of cardboard for the sail; a rocket could be made from cardboard tubes). Now ask if there are any other things they will need to complete the project; have them focus on items to fasten and decorate with. Record their ideas on chart paper.

Have the students use the activity sheet to plan and design their project, and to list the materials they will need for construction.

Once all additional materials have been gathered, set aside an extended period of time for the students to construct their projects and present them to the class.

►

10

Activity Sheet

Directions to students:

Make a design for your special project.
Also list the materials you will need (2.10.1).

Activity Centre

- After completing the projects, you may find that some students will want to build something else based on their classmates' ideas. Provide similar materials at the centre so students can create new projects.

- Keep a big box in the classroom for collecting recycled materials. As an ongoing, year-long activity, have the students use the materials to build items at a construction centre.

Assessment Suggestion

Take photographs of the students with their constructed projects. These can be placed in a science portfolio.

My Construction Project

I am going to build a _____.

I will need these materials:

_____   _____

_____   _____

_____   _____

_____   _____

_____   _____

_____   _____

References for Teachers

Balsamo, Kathy. *Exploring the Lives of Gifted People in the Sciences.* Parsippany, NJ: Good Apple, 1987.

Bosak, Susan. *Science Is... .* Richmond Hill, ON: Scholastic Canada, 1991.

Brecher, Erwin, and Mike Gerrard. *Challenging Science Puzzles.* New York: Sterling Publications, 1997.

Brown, Robert. *200 Illustrated Science Experiments for Children.* Blue Ridge Summit, PA: TAB Books, 1987.

Gardner, Robert. *Science Projects About Kitchen Chemistry.* Springfield, NJ: Enslow, 1999.

Harlow, Rosie, and Gareth Morgan. *175 Amazing Nature Experiments.* New York: Random House, 1992.

Hirschfeld, Robert, and Nancy White. *The Kids' Science Book.* Milwaukee: Gareth Stevens, 1997.

Hutton, Glen. *Household Science.* Parsipanny, NJ: Modern Curriculum Press, 1989.

Lamb, Herb, et al. *Science Today.* Austin, TX: Steck-Vaughn Company, 1987.

Markle, Sandra. *Discovering More Science Secrets.* New York: Scholastic, 1992.

McDonald, Bob, and Eric Grace. *Wonderstruck.* Toronto: Stoddart, 1998.

Murphy, Pat, Ellen Klages, and Linda Shore. *The Science Explorer.* New York: Henry Holt, 1996.

Scott, Corinn Codye. *Chemistry.* Upper Saddle River, NJ: Quercus, 1986.

Smolinsky, Jill, Carol Amato, and Eric Ladizinsky. *50 Nifty Super Science Fair Projects.* Chicago: RGA Publishing Group, 1996.

Suzuki, David. *Looking at the Environment.* Toronto, ON: Stoddart, 1989.

Willow, Diane, and Emily Curran. *Science Sensations.* Reading, MA: Addison-Wesley, 1989.

Unit 3

Energy in Our Lives

Books for Children

Berenstain, Stan, and Jan Berenstain. *The Berenstain Bears' Science Fair*. New York: Random House, 1977.

Berger, Melvin. *All About Electricity*. New York: Scholastic, 1995.

_____. *Switch On, Switch Off*. New York: Crowell, 1989.

Cast, C. Vance. *Where Does Electricity Come From?* Hauppauge, NY: Barron's, 1992.

Cole, Joanna. *The Magic School Bus and the Electric Field*. New York: Scholastic, 1997.

Fowler, Allan. *Energy From the Sun.* New York: Children's Press, 1998.

Glover, David. *Batteries, Bulbs, and Wires.* New York: Kingfisher Books, 1993.

Hewitt, Sally. *Full of Energy.* New York: Children's Press, 1998.

McLellan, Joseph. *Nanabosho Steals Fire.* Winnipeg: Pemmican, 1990.

Thompson, Colin. *The Tower to the Sun.* New York: Knopf, 1996.

Yolen, Jane. *Wings.* San Diego: Harcourt Brace Jovanovich, 1991.

Web Sites

- **http://www.exeloncorp.com/kids/index1.htm**

  Comic characters Volt and Watt take students on a journey of discovery about energy. Keep clicking, and read your way through their exciting missions and learn about energy at the same time. The site also offers students an opportunity to ask Volt and Watt questions, links to other sites, and provides information and activities for parents and teachers.

- **http://www.energy.ca.gov/education/index.html**

  Energy Quest is a great way to learn about how to save energy. This site provides puzzles, resources, projects, and many other options to find out more about scientists, fossil fuels, and energy safety. Also included is a twelve-chapter story, *Devoured by the Dark*, which is a great literature component for this science unit.

- **http://tqjunior.advanced.org/3660/**

  Another great site from Think Quest, which helps students and teachers learn about electronic devices and how they work. Includes stereos, cell phones, calculators, and more. Enter the time machine and learn about the history of technology, discover more about scientists, and find excellent lesson plans and activities on a host of energy topics.

- **http://www.energyed.ergon.com.au/**

  Find out how electricity is made, take a quiz, investigate renewable energy, or do an energy audit at your school. This wonderful Australian site has useful links, a helpful glossary, and much more.

- **http://www.oneworld.org/energy/etree.html**

  A comprehensive look at nonrenewable, renewable, and nuclear energy. Each category includes facts and information about the sources of energy and the environmental impact of energy usage around the world.

- **http://flyhiwaay.net/~palmer/motor.html**

  For those adventurous scientists, an opportunity to build a simple electric motor, using common materials. Step-by-step instructions, with helpful hints and trouble shooting, as well as other interesting links to energy-related information.

- **http://K12.cnidr.org/gsh/schools/ny/che.html**

  Learn about energy uses, past and present, complete with graphs and extensions from Cayuga Heights Elementary School, New York. The information covers energy use in the kitchen, bedroom, and other appliances around the home, with helpful ideas about how to make your house more energy efficient.

- **http://www.newenergy.org/newenergy/**

  Renewable Energy and Sustainable Energy Systems in Canada: a resource site with links to relevant organizations, current news, Internet, and print resources. It also covers the basics of renewable energy sources, including solar power, biomass, and tidal power.

▶

■ **http://www.eskimo.com/~billb/ miscon/whatis.html**

Help dispel the many myths about electricity and energy through the extensive knowledge and information afforded by this site. Also included are scientific articles explaining anything and everything you ever wanted to know about energy.

■ **http://www.swifty.com/apase/ energy.html**

An introduction, history, and discussion of the forms of energy. This site also has a career section, and great activity suggestions designed for students in primary and intermediate grades.

Introduction

Energy has many forms, and it is an integral part of our daily lives. Through investigations, students will understand that they can activate and control the many different forms of energy that they use every day. They will also develop an understanding of the importance of monitoring their energy use. Finally, students will explore and realize that all living things need some form of energy in order to survive.

By the end of this unit, students will demonstrate an understanding of the ways in which energy is used in daily life. They will investigate some common devices and systems that use energy and the ways in which these can be controlled manually. The students will also describe different uses of energy at home, at school, and in the community, and suggest ways in which energy can be conserved.

Science Vocabulary

Throughout this unit, teachers should use, and encourage students to use, vocabulary such as: *energy, movement, muscles, food energy, electricity, propane, gasoline, battery, natural gas, input,* and *output.*

Materials Required for the Unit

Classroom: pencil sharpeners, pencils, chart paper (yellow, dark blue), felt markers, construction paper, tape, scissors, string or yarn, recipe cards, glue, Manila tag paper, poster paper, pencil crayons, masking tape, crayons, paints, paintbrushes

Books, Pictures, and Illustrations: *The Berenstain Bears' Science Fair* (a book by Stan and Jan Berenstain), books, magazines, newspaper flyers, pictures of energy devices (included)

Household: food items (apple, hard-boiled egg, milk)

Equipment: samples of everyday devices that use electrical energy from outlets (e.g., hair dryer, blender, toaster), everyday devices that use electrical energy from batteries (e.g., flashlights, radio)

Other: beanbags, small fan or paper folded into a fan, toy cars or trucks, windup toys, CD (or lyrics or cassette) of the song "Mr. Sun" (by Raffi)

1 **What Is Energy?**

Materials

- *The Berenstain Bears' Science Fair,* a book by Stan and Jan Berenstain
- beanbags
- small fan or paper folded into a fan
- toy cars or trucks
- pencil sharpeners
- pencils
- books
- windup toys
- chart paper, felt markers

Activity

Begin the lesson by reading the book *The Berenstain Bears' Science Fair.* Through discussion, focus on the term *energy* as introduced in the book.

Have the students sit in a circle. As a group, do several tasks that show forms of energy:

- Throw a beanbag to a friend.
- Fold a piece of paper and fan yourself.
- Push a toy car or truck across the floor.
- Sharpen a pencil.
- Lift a book.
- Jump on the spot.
- Wind up a toy and let it move across the floor.

Once the students have completed all of the tasks, ask them:

- How were all of these activities the same? (all involved movement)

Demonstrate each of the tasks again. Ask the students:

- What is happening when I am doing this task?
- What is moving?
- What made it move?

Title a sheet of chart paper with the heading Energy. Divide the sheet into two columns labelled Things That Move and How They Move. Have the students brainstorm a list of things that move (e.g., humans, car, fan, clouds, swing, bird, bike). Beside each item, record what made the object move (e.g., energy, muscles, gas, wind).

Referring back to the book *The Berenstain Bears' Science Fair,* as well as to the activities that the students did, ask:

- What do you think energy is?

Record their definitions of energy on chart paper. Keep their definitions of energy posted in the classroom throughout the unit. Students may find they wish to add to or alter their definitions as they gain more knowledge on the topic.

Activity Sheet

Directions to students:

Look at each picture carefully. Circle the moving thing in each picture. Below the picture, print what is moving and tell what is making it move (3.1.1).

Extensions

- Have students draw pictures of different items in their home that move. Beside each item, have the students describe what moves and what makes the object move.

- Walk around the school and the schoolyard. Have students identify moving objects and take photographs of the objects. Create a picture chart by placing the photographs on chart paper and labelling each object. Have the students identify what made each object move and record their ideas beside each photograph.

- Learn the lyrics to *The Marvelous Toy,* a song by Fred Penner, to further examine how objects move.

What Is Energy?

What is moving? _____

Source of energy: _____

What is moving? _____

Source of energy: _____

What is moving? _____

Source of energy: _____

What is moving? _____

Source of energy: _____

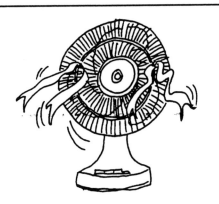

What is moving? _____

Source of energy: _____

What is moving? _____

Source of energy: _____

2 | Energy From the Sun

Materials

- lyrics, cassette, or CD of the song "Mr. Sun" by Raffi
- chart paper, felt pens
- yellow and dark blue construction paper
- tape
- scissors

Activity

As an introduction to the lesson, teach the students the song "Mr. Sun." Title a sheet of chart paper with the heading What We Know About the Sun. Have the students share what they know about the Sun, and record these ideas on the chart paper.

Using yellow construction paper, have the students draw and cut out a sun, then tape it onto the dark blue construction paper. Explain to the students that you are going to place their sun pictures in a window for several days (with the suns facing outward). Ask:

- Do you think anything will happen to your pictures?
- What do you think might happen? Why?

Tape the pictures in a window that gets a lot of sunlight. Several days later, remove the pictures from the window and pass them out to the students. Have the students closely observe the pictures. Ask:

- Has anything changed about your picture?

Have the students carefully remove the suns from their blue construction paper and examine the blue paper. Ask:

- What happened to the blue construction paper?
- What part of the paper is faded?
- Why do you think this happened?
- What role did the Sun play in this?

- What would have happened if the coloured paper had been left in front of the window for a longer period of time?

Explain to the students that energy from the Sun bleached the paper. Have the students look back to their chart paper definition of energy (see page 138). Ask them:

- Can you think of something you should add to your definition of energy?
- What did the energy from the Sun do to the paper?

Discuss the Sun further. Add to the chart more things that the students know about the Sun.

Examples may include:

- The Sun lets us see things.
- The Sun keeps us warm.
- The Sun gives us daytime.
- The Sun can help dry clothes on a clothesline.
- The Sun can give you a sunburn.
- The Sun can be dangerous to your eyes, so you should not look directly at it.

Once you have created and reviewed your list, ask the students:

- What would happen if there were no Sun?
- What would happen to plants?
- What would happen to animals?
- What would happen to humans?
- Could humans survive without the Sun?
- What does this tell you about the Sun?

Reinforce to the students that the Sun is the most important source of energy on Earth. Without it, living things could not survive.

Activity Sheet

Directions to students:

Use the chart to draw diagrams of things you know about the Sun's energy (3.2.1).

▶

2

Extensions

■ Compile the students' activity sheets into a class book titled *What We Know About the Sun*. Place the book in the classroom library or share it with other classes.

■ Conduct a classroom experiment on the effect of sunlight on the growth of plants. Place one plant in front of the window. Water and fertilize it as required. Place a second plant in a cupboard or dark corner of the classroom. Water and fertilize it as required. Have the students record the growth of the plants over a period of time. You may wish to graph the results.

■ Fill two bowls with water. Place one bowl in direct sunlight; place the other bowl in a dark corner of the classroom. Put a thermometer in each bowl and observe and compare changes in water temperature.

■ Place one tray of frozen ice cubes in the direct sunlight and another tray of frozen ice cubes in a dark corner of the classroom. Observe which tray of ice cubes melts first. Why?

Assessment Suggestion

Observe students as they discuss what they know about the Sun, examine and describe their sun pictures, and as they provide ideas about the Sun as a form of energy. Use the anecdotal record sheet on page 15 to record results.

The Sun's Energy

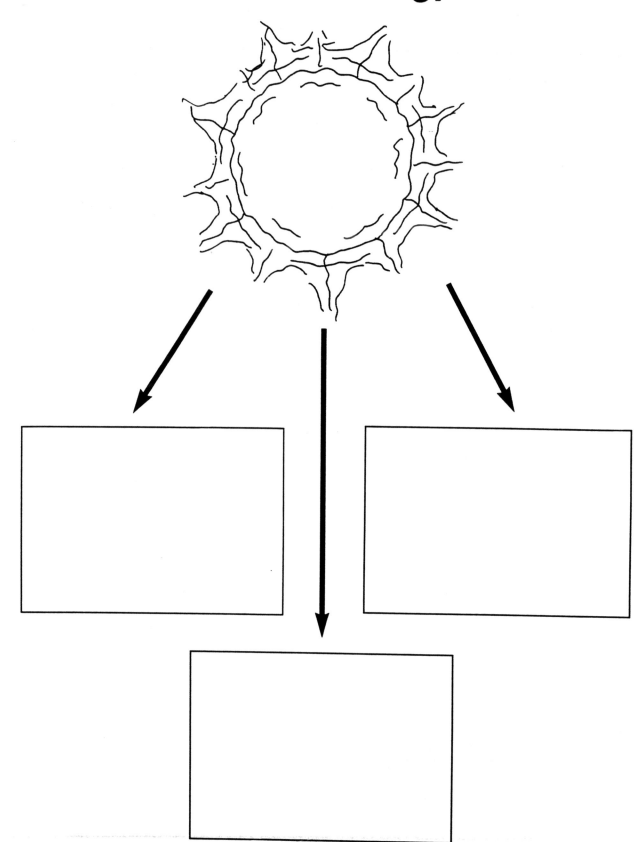

3 Living Things and Energy

Materials

- string or yarn
- scissors
- recipe cards
- felt markers
- food items, such as a hard-boiled egg, an apple, a glass of milk
- chart paper
- masking tape

Activity

Review with the students the importance of the Sun as the principal source of energy for Earth. Now ask the students:

- Where do we get our energy from?
- What do you feel like when you do not have energy?
- How can you get more energy?

Explain to the students that food is the primary source of energy for humans and all living things. Tell the students that they are going to help you trace food energy for humans back to the Sun.

Have the students sit in a large circle. Select one student to stand near the inside of the circle. Give the student a food item (e.g., a hard-boiled egg). Print "egg" on a recipe card and stick it with masking tape to the front of the student's shirt. Give the student a piece of string to hold on to. Now ask the rest of the students:

- Where did the egg come from?

Have another student stand up and enter the circle. Print "chicken" on a recipe card and stick it with masking tape on the front of the student's shirt. Have the student take the other end of the string and hold on to it. This will show the connection between the egg and the chicken. Now ask the rest of the students:

- Where does the chicken get its energy from?

Have a third student stand up and enter the circle. Write "grains" on a recipe card and stick it on the front of the student's shirt with masking tape. Cut an additional piece of string. Have the "chicken" and the "grain" students hold on to either end of the string. The students will be creating a concrete link of energy by using the string. Now ask the rest of the students:

- Where does the grain get its energy from to grow?

Have a fourth student stand up and enter the circle. Write "Sun" on a recipe card and stick it on the front of the student's shirt with masking tape. Cut a third piece of string. Have the "grain" and the "Sun" students hold on to either end of the string. This will complete the entire link from the egg to the Sun.

You may wish to illustrate the same connections back to the Sun for other food items such as an apple or a glass of milk.

Title a sheet of chart paper Energy for Living Things. Divide the chart into two columns, titled Living Thing and Food Energy. As a class, brainstorm a list of living things and record these ideas in the first column of the chart. In the second column of the chart, identify the food sources for each living thing. Examples may include:

- robin and worms
- human and meat (vegetables, fruit, grains, dairy products)
- fish and minnows
- rabbit and carrots
- flower and energy from the Sun

▶

3

Note: Students will need to thoroughly discuss the concept that plants get their food energy from the Sun. At this grade level, it is not necessary or appropriate to go into great detail about the process of photosynthesis and the production of chlorophyll. Instead, simply focus on the idea that plants make their own food from the energy they get from the Sun.

Activity Sheet

Directions to students:

On the chart, draw diagrams of living things. Print the name of each living thing below your diagrams. Now draw diagrams to show how each living thing gets its energy from food (3.3.1).

Extensions

■ Visit a local food processing plant. Observe, firsthand, the ingredients and methods for making various food items.

■ Ask a public health nurse to come into the classroom and talk to the students about good nutrition and a balanced diet.

Energy for Living Things

Living Thing	Food Energy
_____	_____
_____	_____
_____	_____

4 Everyday Uses of Energy

Materials

- charts made in previous lessons, defining energy, identifying forms of energy
- chart paper, felt pens
- masking tape
- magazines, catalogues, newspaper flyers
- scissors, glue
- samples of everyday devices (e.g., appliances such as a hair dryer, blender, and toaster) that use electrical energy from outlets; as well as items that use electrical energy from batteries (such as flashlights and radios)

Activity

Before starting this activity, have the students review the charts that define energy and give examples of energy. Discuss the idea that energy comes in many different forms, such as energy from the Sun and energy from food. Focus on the previously made chart that identifies objects that move and their forms of energy (see page 138). Select an example that discusses gasoline energy. Ask:

- How does the car get its energy?
- What other objects use gasoline for energy and movement?

Title a sheet of chart paper Everyday Uses of Energy and divide the chart into two columns. Title the first column Type of Energy and the second column Example. Under the first column, record the term *gasoline* and then in the second column list the students' responses (e.g., motorcycle, boat motor, lawn mower, snow blower, grass trimmer, bus, van).

Explain to the students that there are other forms of gas that many of us use every day. Natural gas is used in many furnaces to heat our homes. Propane gas is used for many barbecues. Add these ideas to the chart.

Now display an electrical appliance such as a hair dryer or toaster. Have students examine the cord and plug on the appliance. Ask:

- How does this object work?
- What type of energy does it use?
- Where does the electricity come from?

Record the term *electricity* in the first column on the chart. Ask:

- What other objects use electricity from outlets?

Record their responses in the second column on the chart.

Display an example of an item that uses electrical energy from batteries, such as a flashlight. Have students examine the flashlight and take it apart to view the batteries inside. Ask:

- How does this object work?
- Where does it get its energy from to light up?

Record the term *batteries* in the first column on the chart. Ask:

- What other objects use batteries for energy?

Record their responses in the second column on the chart.

Activity Sheet

Directions to students:

Use the magazines, catalogues, and flyers to find examples of everyday devices that use energy from electrical outlets and from batteries. Cut out the examples, sort them, and glue them onto the Venn diagram (3.4.1).

Name: _____

Date: _____

Energy Devices

Energy From Batteries

Energy From Electrical Outlets

5 Safety With Everyday Devices

Materials

- pictures of different energy devices (included) (3.5.1-3.5.6)
- poster paper
- art supplies such as paint, paintbrushes, markers, crayons, pencil crayons

Activity

Have the students sit in a circle. Hold up each of the pictures illustrating situations where humans are using everyday devices. Have the students describe each picture. Make sure they identify the energy-using device in each picture and the type of energy used.

Picture 1: children on a school bus

Picture 2: family barbecuing with a propane barbecue

Picture 3: children working at a computer

Picture 4: adult making toast

Picture 5: adult boiling water in an electric kettle

Picture 6: child playing with a remote-control car

As each picture is introduced, focus discussion on ways that humans stay safe while using the device, as well as on how our senses can help us stay safe. For example:

- What do you have to remember to stay safe on the school bus?
- What should passengers do to stay safe?
- What should the driver do to stay safe?
- How does your sense of sight help you stay safe on a school bus?
- How does your sense of hearing help you stay safe on a school bus?

Continue this process for each device. Also encourage students to provide other examples of safety, such as not sitting too close to a television, keeping electrical cords out of traffic areas, not shining flashlights into someone's eyes, never using the oven or stove without adult supervision, and so on. For each example, also focus on how the senses help us stay safe while using the devices.

Challenge the students to create posters depicting examples of how to stay safe while using energy devices. They can use the activity sheet to plan and design the posters. Have them present their posters to the class and then display them around the school.

Activity Sheet

Directions to students:

Plan your poster to show an example of how to stay safe while using energy devices. Check your spelling, printing, and diagrams before starting your final poster (3.5.7).

Extensions

- View movies and videos about traffic safety, or invite a police officer to present to class on this topic.

- Invite a local firefighter to present to the class on how to safely use energy devices in the home and school to avoid fires.

- Invite a guest from your local hydroelectric company to discuss safety with electrical devices.

5

Assessment Suggestion

Identify criteria for the students' safety posters.
These may include:

- appropriate title
- clear illustrations
- identified energy device
- identified safety issue
- clear presentation

List these criteria on the rubric on page 19.
Observe the students as they present their
posters to the class, and record results.

Date: _____ Name: _____

Designing a Safety Poster

Check ✔ ☐ **Diagrams** ☐ **Printing** ☐ **Spelling**

6 Loss of Energy

Materials

- chart paper
- markers

Activity

Review electrical energy, focusing on the many items in a home that use electricity. Now have the students close their eyes and listen as you tell a story of what could happen if there was no electricity.

Have the students imagine they are at home with their families on a spring evening. There are reports on the television and radio of a heavy thunderstorm on its way. Soon, the winds blow strong, rain pours from the sky, and hail begins to pelt the ground. All of a sudden the house goes silent and black. They look outside to see that trees have fallen all around and have broken the hydro wires that bring electricity to their house.

Now have the students open their eyes and discuss what they will do. Have the students brainstorm titles for their imaginary storm. As a class, select a title and print it on chart paper. Divide the chart into two columns; title one column Problem and the other column Solution. Ask the students:

- What problems do you encounter when you lose power in your home? (e.g., television does not work, lights do not work, electric oven does not work, appliances do not work, there is no heat for the house)

Record each of their suggestions under the problem heading. Now invite the students to find solutions to these problems. Ask them:

- How could you and your family find solutions to these problems until the hydro lines were repaired?

Review each problem and in the second column of the chart paper, write solutions.

For example:

Problem	Solution
■ no television	read a book, do a puzzle, play outside
■ no lights	use candles to provide light
■ no oven	cook meals on a barbecue
■ no heat	use a fireplace for heat

Record additional problems and solutions discussed by the class. Now have the students use the activity sheet to draw an illustration of their experiences during the imaginary storm.

Activity Sheet

Directions to students:

Draw a diagram showing you and your family during the imaginary storm. Under your diagram, explain how you and your family might spend the evening with no electricity (3.6.1).

Extensions

- Research how people lived prior to the discovery of electricity (e.g., What was everyday life like for a pioneer?). Culminate the research with a "Pioneer Day" at school.

- Read books about how electricity was first discovered.

- Focus on energy conservation and discuss the importance of reducing the amount of electricity we use. Have the students design a poster on conserving energy at home or school. Display the posters throughout the school or have the students present their "Conserving Energy" posters to other classes in the school.

6

- As a class, prepare a write-up for the school newsletter on how to conserve energy at home, at school, and at work.

- Invite an employee from the local hydro company to discuss the importance of energy conservation.

Date: _____   **Name:** _____

Our Imaginary Storm

_____

_____

_____

_____

_____

References for Teachers

Norris, Jill. *Energy: Light, Heat, and Sound*. Monterey, CA: Evan-Morr Corp., 1998.

Ticotsky, Alan. *Who Says You Can't Teach Elementary Science*. Glenview: Good Year Books, 1985.

Unit 4
Everyday Structures

Books for Children

Blegvad, Erik. *The Three Little Pigs*. New York: Atheneum, 1990.

Bryant-Mole, Karen. *Tools*. Parsippany, NJ: Silver Press, 1997.

Byars, Betsy. *Beans on the Roof.* New York: Delacorte, 1988.

Carle, Eric. *Papa, Please Get the Moon for Me*. Saxonville, MA: Picture Book Studio, 1991.

Gibbons, Gail. *How a House Is Built*. New York: Holiday House, 1990.

_____. *Tool Book*. New York: Holiday House, 1982.

Hoberman. Mary Ann. *A House Is a House for Me*. New York: Puffin, 1982.

Jessop, Joanne. *Big Buildings of the Ancient World*. New York: F. Watts, 1993.

_____. *Big Buildings of the Modern World*. New York: F. Watts, 1994.

Johmann, Carol. *Bridges!: Amazing Structures to Design, Build & Test*. Charlotte, VT: Williamson, 1999.

MacAulay, David. *Unbuilding*. Boston: Houghton Mifflin, 1980.

Madgwick, Wendy. *Super Materials*. Austin, TX.: Raintree Steck-Vaughn, 1999.

Massi, Jeri. *The Bridge*. Greenville, SC: Bob Jones University Press, 1986.

Morris, Ann. *Houses and Homes*. New York: Lothrop, Lee & Shepard Books, 1992.

Neville, Emily Cheney. *The Bridge*. New York: Harper & Row, 1988.

Shulevitz, Uri. *The Secret Room*. New York: Farrar, Straus & Giroux, 1993.

Sturges, Philemon. *Bridges Are to Cross*. New York: G.P. Putnam's Sons, 1998.

Wheatley, Nadia, and Donna Rawlins. *My Place*. Brooklyn, NY: Kane/Miller Book Publishers, 1992.

Wilkinson, Philip. *Amazing Buildings*. New York: Dorling Kindersley, 1993.

Winch, Madeleine. *Come By Chance*. New York: Crown Publishers, 1988.

Web Sites

- **http://www.swifty.com/apase/charlotte**

  Association for the Promotion and Advancement of Social Education: an excellent resource for teachers and students. Click on "Charlotte's Web" for educational ideas on human environments – includes architecture for kids, rhythm and shape, urban design, and much more.

- **http://www.field-guides.com/trials.htm**

  Take an extraordinary field trip to such places as the Natural Wonders of the World and discover, among other things, how they were formed. Each trip begins with teacher resources and necessary background information, including terms and concepts that are particular to the chosen area. All sites include visuals, legends, and links.

- **http://www.yesmag.bc.ca/focus/ structures/structures.html**

  Canada's Science Magazine for Kids: click on "Focus On" underneath the heading "Departments." This will take you to Structures, an excellent site for students and teachers focusing on human-made and natural structures, including bird nests, ant hills, beaver dams, and spider webs.

- **http://www.library.advanced.org/18788/**

  Architecture Through the Ages: an educational site that discusses all types of architecture from all around the world and from all time periods. Excellent site for teachers and students researching architecture.

- **http://www.endex.com/gf/buildings/ ltpisa.html**

  Leaning Tower of Pisa: scroll down for history and information on the town of Pisa and its architecture (including the Tower of Pisa). Includes photographs and other visuals.

- **http://www.ScienceU.com/**

  Learn more about geometry at the geometry centre: includes interactive activities (with quicktime VR), activities for the classroom, and facts and figures. This site also includes an observatory and library.

- **http://cs.eng.usf.edu/pharos/ wonders/other.html**

  Explore the forgotten, modern, natural, and seven ancient wonders of the world. Each page offers pictures, information, and maps, as well as great links to lead you to more in-depth knowledge.

- **http://burgoyne.com/pages/bldgconn/ a.htm**

  Learn more about the history of bungalows. This site provides floor plans and an activity that involves building your own model bungalow.

Introduction

In this unit, students will observe, classify, and manipulate a wide variety of objects and structures in natural and human-made environments. The objects and structures will have distinctive shapes, patterns, and purposes. Students will begin to identify shapes that are repeated in various patterns (e.g., square, triangle, circle) and shapes and patterns that are common to most structures.

By the end of this unit, students will demonstrate awareness that structures have distinctive characteristics. They will design and make structures that meet a specific need.

The students will also demonstrate an understanding of the characteristics of different structures and ways in which the structures are made.

Note: The concept of structures may seem difficult for young children to understand unless you can provide examples from the students' own immediate environment. It is important, therefore, to present this unit in a real-life context, focusing on structures that students see and use every day.

Students will also be introduced to the concept of a system. They will observe and use systems they encounter in everyday life that involve a single input (the actions required to set a system in operation) and a single output (the response of the system): for example, the flicking of a light switch (input) causes the light to illuminate (output).

Science Vocabulary

Throughout this unit, teachers should use, and encourage students to use, vocabulary such as: *structure, pattern, shape, furniture, buildings, human, natural, tool, system, input,* and *output.*

Materials Required for the Unit

Classroom: chart paper, markers, Plasticine, masking tape, paint, stickers, mural paper, scissors, glue, Play-Doh, modelling clay, rulers, hole punch, stapler, construction paper, felt pens, paintbrushes

Books, Pictures, and Illustrations: *A House Is a House for Me* (a book by Mary Ann Hoberman), pictures of various pieces of furniture (included), pictures of natural structures and human-made structures (included)

Household: measuring spoons, spatula, forks, can opener, whisk, jewellery, dolls, Ziplock bags

Equipment: Polaroid camera (and film)

Other: cardboard tubing (from paper towels, toilet paper), toothpicks, straws, pipe cleaners, Hula-Hoops, small cardboard boxes (and other building material for constructing buildings), natural materials for building bird nests, hammers, wrenches, pliers, screwdrivers, building blocks, Lego, pattern blocks, hockey cards, photographs, bags, cloth, fasteners (buttons, string, togs), Velcro, sparkles, sequins, toy trucks and cars (that can be taken apart)

1 Playground Structures

Materials

- chart paper
- markers
- variety of materials for students to use in constructing model playground structures (e.g., cardboard tubing from wrapping paper, toilet paper, and paper towels; tape, toothpicks, and Plasticine; straws, pipe cleaners)
- materials for finishing or decorating playground structures (e.g., paint, paintbrushes, stickers, markers)
- camera for taking pictures of the playground structures (optional)

Activity

Note: Access to a school or park playground structure is necessary for this lesson.

Take the students to a school or park playground. Have them observe the various structures in the playground. Allow the students time to play on the equipment. If you have a camera, take pictures of the students as they play on the structures.

Following the play time, group the students together for discussion. Explain that each part of the playground equipment is called a structure. Ask:

- What different structures did you play on? (Some answers might be: ladder, slide, monkey bars, swing, and ramp.)
- What are the structures made of?
- Which is the biggest structure?
- Which is the smallest structure?
- What shapes do you see on the structures? (Some responses might include: triangle, square, rectangle, and circle.)
- Do you see any patterns on the structures? (Some responses might include: repeating squares on a ladder and repeating boards on a walkway or ramp.)

- How are all the structures the same?
- How are they different?

Back in the classroom, have the students make a list of all the structures they played on. Record their ideas on chart paper. Use the photographs or their observations to discuss the structures in terms of materials, size, shapes, patterns, and functions.

In pairs, have the students select a playground structure to design and construct. Record the selected structures on chart paper. As a class, brainstorm a list of possible materials that the students could use to build their structures. Encourage them to plan their design by drawing a diagram of the structure using the activity sheet. Display a wide variety of construction materials for building the structures, and have the students identify the materials that they could use to build the structures. Encourage them to think about how they will fasten together the parts of their structure (e.g., tape, Plasticine, glue). Have the students list these materials on their activity sheet.

Provide plenty of time for students to manipulate materials and construct their playground structures. When they are completed, display the structures for all students to examine.

Activity Sheet

Directions to students:

Choose a structure from the playground to build. Print the name of your structure on your sheet and draw a diagram of it. List the materials you will need to build your structure (4.1.1).

▶

1

Extensions

- Create a Super Playground: Display all of the structures built by the students.

- Visit other parks or playgrounds to draw comparisons of different types of playground structures.

- Classify playground structures according to the materials they are made of (e.g., wood, metal, rubber, and plastic).

Activity Centre

Provide additional construction material and allow students time to build other playground structures.

Assessment Suggestion

Observe students as they design their playground structures, identify materials, and construct the structures. Use the anecdotal record sheet on page 15 to record results.

Playground Structures

I will build a _____.

I will use these materials:

_____   _____

_____   _____

_____   _____

_____   _____

2 | Structures in the School

Materials

- chart paper
- markers
- Hula-Hoops
- Polaroid camera and film
- pictures of various pieces of furniture, such as a chair, desk, stool, bookshelf, and filing cabinet (included) (4.2.1)

Note: If you do not have access to a Polaroid camera, pictures of classroom furniture can be cut out of school and office supply catalogues. These can then be used for the sorting activity with Hula-Hoops.

Activity

Review the previous activity on building playground equipment. Ask:

- What do you call the different types of playground equipment? (structures)
- Do you think that there are other types of structures around you right now?

Display a piece of classroom furniture, such as a student's chair. Ask:

- What is this called?
- What is a chair used for?
- How is this chair different from the playground structures?
- How is this chair the same as the playground structures?

Explain to the students that the chair is also a structure, as is all furniture. Have the students look around the classroom and identify all the different pieces of furniture they see. Record these on chart paper. Discuss the various pieces of furniture. Ask:

- What are they made of?
- What shapes do you see in the structures?
- Which structure is the biggest?
- Which is the smallest?

Now take the students on a "structure parade" around the school to identify other furniture structures. Visit other classrooms, the library, the office, the computer lab, and the gym to see how many different structures the students can find. As you parade around, stop to discuss the characteristics of various pieces of furniture, focusing on the design, materials, shapes, patterns, size, and so on of each piece. Also take photographs of as many of the furniture structures as is possible.

Back in the classroom, display the photographs for students to discuss. Play the sorting game What Is My Rule? using the photographs and Hula-Hoops. Place several of the photographs that have been classified inside one of the two Hula-Hoops, according to your own predetermined rule (e.g., furniture with legs and furniture with no legs). Challenge the students to identify your sorting rule. Continue this game by classifying the pictures according to different rules such as:

- metal furniture and wood furniture
- big furniture and small furniture
- furniture to sit on and furniture to stand on

Also challenge the students to sort the photos in different ways and have their classmates identify the sorting rules.

Activity Sheet

Note: The Venn diagram can be enlarged onto bigger paper so there is space for sorting pictures.

Directions to students:

Cut out the pictures of the furniture and use the Venn diagram to sort your pictures (4.2.1).

▶

2

Extensions

■ Discuss furniture found in homes and compare it to furniture found in a school. With the help of family members, have students record the names, or draw diagrams, of furniture found in different rooms in their homes.

■ Read the story *Goldilocks and The Three Bears* and discuss the characteristics of the furniture in the story.

Activity Centre

■ Have the students build a doll house from a large cardboard box, then use smaller boxes and other materials to build furniture for the house.

■ Have the students cut out pictures from catalogues of furniture for the home. Using the activity centre sheet as a "blueprint," have the students glue the furniture pictures into the various rooms (4.2.2).

Note: Enlarge the activity centre sheet to 11" x 17" for student use.

Sorting Furniture

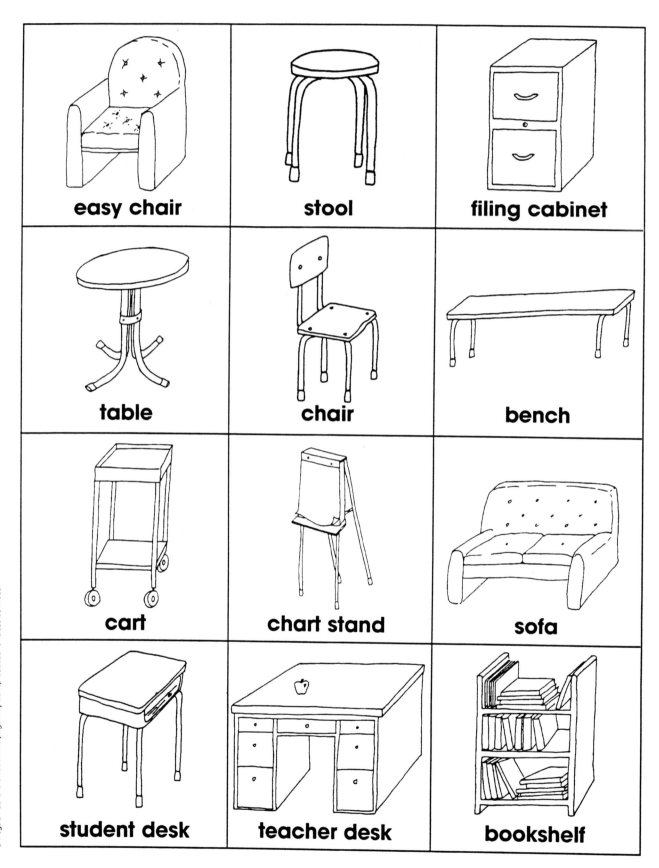

easy chair	stool	filing cabinet
table	chair	bench
cart	chart stand	sofa
student desk	teacher desk	bookshelf

Sorting Furniture

Furniture in a Home

Bedroom	Bedroom	Bathroom
Kitchen	Living Room	

3 Buildings as Structures

Materials

- *A House Is a House for Me*, a book by Mary Ann Hoberman
- chart paper
- markers
- small cardboard boxes and other building material for constructing buildings
- mural paper

Activity

Read the book *A House Is a House for Me* aloud. As you read, discuss various ideas in the book. Ask the students:

- Is a house a structure?
- What other buildings are shown in the book?

After reading and discussing the book, have students brainstorm a list of all the types of buildings they might find in their community. Record these ideas on chart paper.

Go for a walk around the community to identify various buildings. Encourage students to observe carefully and then discuss the features of the buildings they see. Ask:

- What are the buildings made of?
- What different parts of the buildings can you see? (stairs, doors, doorknobs, mailboxes, windows, roof, chimney, and so on)
- Which are the largest buildings?
- Which are the smallest buildings?

Back in the classroom, have the students add the buildings they saw in the community to their list. Also discuss further the parts of the buildings they saw.

Challenge the students to construct their own community by designing buildings. Set out a large sheet of mural paper on a large table or cleared section of the classroom. On the paper, have the students plan their community, complete with streets, parks, and buildings. While students are working, engage them in discussions about the structures and their functions. Have them use the cardboard boxes and other construction materials to design and build several buildings for their community. Encourage the students to include residential streets, as well as streets with stores, businesses, and other buildings in their plans.

Activity Sheet

Directions to students:

On the house, draw all the parts that are missing. Label each part. You may also add other structures that might be found around the house (e.g., garage, doghouse, swing set, fence, patio, deck) (4.3.1).

Assessment Suggestion

Have students complete a science journal sheet on page 17 to reflect on what they have learned about buildings as structures.

Name: _____

Date: _____

What Is Missing?

4 Natural Structures

Materials

- *A House Is a House for Me*, a book by Mary Ann Hoberman
- scissors, glue
- chart paper, felt pens
- natural materials for building bird nests (students can collect these)
- Plasticine, Play-Doh, or modelling clay
- pictures of natural structures (bird nest, honeycomb, spider web, beaver dam) and human-made structures (apartment building, stool, fence, bridge) (included) (4.4.1)

Activity

Re-read the book *A House Is a House for Me*. Focus on the structures introduced in the book. Ask the students:

- Which structures are built by humans?
- Which structures are not built by humans?

Record the names of several natural structures, such as an anthill, beehive, spider web, bird nest, seashell, and so on. If available, display samples or pictures of some of these items for students to observe and examine. Ask:

- What is a bird nest made from?
- How do you think the bird makes the nest?

Over the next few days, have students collect natural materials that they could use to build a bird nest (e.g., twigs, packing straw, grass). Have them construct nests and compare them to an actual bird nest. Students can also make birds for their nests, using Plasticine, Play-Doh, or modelling clay.

Note: Natural materials such as twigs and grass can be moulded together using diluted white glue.

Activity Sheet

Directions to students:

Cut out the pictures and sort them into two groups, one showing structures made by humans and one showing natural structures. Glue the pictures under the correct headings on the chart (4.4.1)

Extensions

- Have students re-create spider web patterns. On black construction paper, have them first draw a spider web, then trace the pattern with white glue. Pour salt over the entire web pattern, then shake the excess salt into a bowl. Students can then construct spiders from paper, clay, or model magic to attach to their webs.

- Have students re-create honeycomb patterns using cut-out sections from cardboard egg cartons.

Activity Centre

Encourage students to bring in examples of different natural structures. Display the structures on a table. Include magnifiers so that students can closely examine the natural structures. Review handling procedures before students visit the activity centre.

Structures

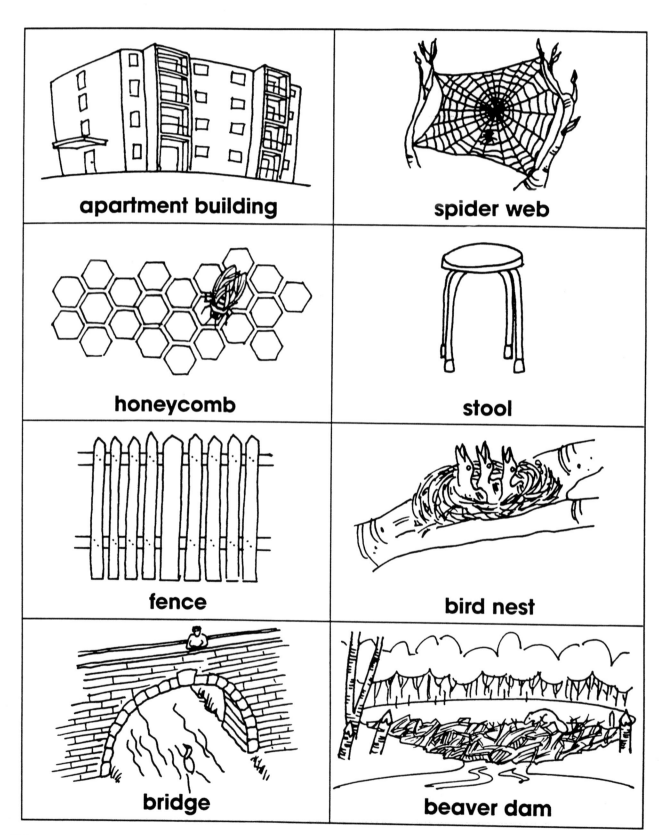

apartment building

spider web

honeycomb

stool

fence

bird nest

bridge

beaver dam

Date: _____ Name: _____

Structures

Structures Made By Humans	Natural Structures

5 Useful Tools

Materials

- classroom tools such as scissors, rulers, hole punch, stapler
- kitchen tools such as measuring spoons, forks, can opener, whisk, spatula
- construction tools such as hammers, wrenches, pliers, screwdrivers
- Hula-Hoops
- additional tools that students bring from home

Activity

Display all the tools for students to examine and discuss. As each tool is focused upon, ask:

- What is this tool called?
- How is this tool used?

Have students demonstrate how several of the tools are used.

Now challenge the students to sort the tools in different ways and place them in the Hula-Hoops according to their rule. Tools can be sorted according to where they are used (e.g., classroom, kitchen, workshop), their function (e.g., putting together, taking apart, mixing, cutting), or in other ways such as by size, shape, colour, material, and so on.

Have each student bring a tool from home that he or she presents to the class through a

project. Encourage the students to bring in a tool different from those used in class. (You may wish to have additional tools on hand for students to use for this project.)

Using the activity sheet, have students identify the tool, draw a diagram of the tool, and explain how it is used. Students can then present their tools to the class. After presentations, the activity sheets can be put together in the form of a book and displayed along with the various tools.

Activity Sheet

Directions to students:

Draw a picture of your tool and explain how you would use it (4.5.1).

Extensions

- Invite a carpenter to the class to demonstrate how tools are used to build things.

- Involve the students in several cooking activities to give them firsthand experience using measuring tools and kitchen utensils. During these activities, challenge students to select appropriate tools for a specific job (e.g., What would you use to scrape cookie dough from the bowl?).

Date: _____   Name: _____

Useful Tools

My tool is a _____.

This is how my tool is used:

_____

_____

_____

_____

6 Designing and Constructing Structures

Materials

- large collection of building blocks, pattern blocks, Lego, and so on
- large collection of objects such as hockey cards, jewellery, dolls, books, spoons, photographs (or possibly something that you collect personally)
- chart paper
- markers
- variety of materials to make container structures (e.g., boxes, bags, cloth)
- variety of materials to use for fasteners (e.g., buttons, string, pipe cleaners, togs)
- Velcro
- materials to decorate the containers, such as paint, construction paper, glue, sparkles, sequins, stickers, and so on

Activity

Prior to the lesson, dump the collection of blocks on the floor. Have the students form a circle around the blocks. Challenge a student to move the entire collection of blocks out of the circle in one trip. Ask:

- Why is it difficult for you to move the blocks?
- How could you solve this problem?
- What could you use to carry the blocks in?

Brainstorm the names of a variety of containers that the blocks could be placed in to move them more easily (e.g., tray, box, shopping bag, knapsack).

Display one of the collections for the students. Explain that this group of objects is called a *collection*, because someone has collected the objects as a hobby. Explain that some humans collect coins, stamps, and rocks. Ask the students:

- What other things might humans collect?
- Do you have a collection of your own?

As students respond, record on chart paper the names of various collections that they and others collect. Over the next week or two, encourage students to bring in a collection of their own. This could include a collection of toys, dolls, rocks, cars, cards, pennies, or any other items of the students' choice.

Once students have brought in collections, discuss their objects, as well as the containers in which they brought their collections to school. Now challenge the students to make a "treasure chest" in which to keep their collection safe. Provide the following guidelines:

1. Your container should be large enough to hold your whole collection.
2. Your container should be able to be closed and fastened so that your collection will not fall out.
3. Your container should be decorated.

Have students select materials to construct their treasure chests. This may be as simple as a box that they decorate and tie with a fancy ribbon. They may also choose to add special fasteners to keep the lid of the box securely closed (e.g., pipe cleaners, buttons and string, togs).

Once students have made their treasure chests, have them complete the activity sheet, then present their final products to the class, along with their collection.

▶

6

Activity Sheet

Directions to students:

Draw a diagram of your completed treasure chest. List the materials you used to build your chest (4.6.1).

Extension

Focus on the use of fasteners on clothing. Display clothing items with buttons, zippers, Velcro closures, laces, hook and eye fasteners, snap domes, and so on for students to examine and manipulate.

Assessment Suggestion

Identify criteria for the construction of the treasure chests. These might include:

1. appropriate materials selected
2. project completed
3. size is appropriate to fit collection
4. chest can be closed and fastened
5. chest is decorated

List these criteria on the rubric on page 19 and record results for each student.

My Treasure Chest

My treasure chest will hold _____.

This is what I used to build my treasure chest:

_____   _____

_____   _____

_____   _____

_____   _____

7 | How Systems Work

Materials

- chart paper (divided in half: label one column Input and the second column Output)
- felt pens
- collection of toy trucks and cars that can be taken apart to examine their operating systems
- tools to use for taking apart the systems (screwdrivers, pliers, and so on)
- Ziplock bags

Activity: Part One

As you begin the lesson, turn out the classroom lights. Ask the students:

- What just happened?
- Why did the lights go out?
- What did I have to do to make the lights go out?

Explain to the students that your action, or input (flicking the light switch down), caused a response, or output (the lights went out). Under the Input column on the chart paper print "flicking the light switch down." Under the Output column print "lights went out." Ask:

- What input would make the light go back on?

Have a student demonstrate the input and output and add these to both columns of the chart.

Have the students think of other everyday systems that have an input and an output. You may need to encourage their ideas by providing the names of some of these operating systems. Some suggestions include:

- Press the handle on the toilet, the toilet flushes.
- Turn the switch on the computer, the computer starts.

- Press a button to ring a doorbell, the doorbell rings.
- Press the button on an icemaker, ice falls out.
- Press the button on a vending machine, a drink or candy falls out.
- Press the lever on a toaster, the toaster heats up.
- Push the button on the remote, the television turns on.
- Turn the key in the car ignition, the engine starts.

Record the students' suggestions on the chart paper under the appropriate columns.

Activity: Part Two

Select a sample toy car or truck. Pass the toy around so students can examine and manipulate the parts. Ask:

- Which parts of the car move?
- How are the wheels attached to the car?

Introduce the term *axle* and have students identify the axle on several cars and trucks. Examine and identify other parts of the toys such as the steering wheel, windshield, doors, trunk, hood, and so on.

Divide the class into working groups and provide each group with a variety of cars and trucks, tools, Ziplock bags, activity sheets, and pencils. Have the students select a toy and draw a diagram of it on the activity sheet. Now challenge them to take the toys apart, identify the parts, and draw and label each part on the activity sheet. Encourage the students to use the Ziplock bags to keep the small parts in. You may also want to challenge the students to put the cars and trucks back together again.

▶

7

Activity Sheet

Directions to students:

Draw a diagram of your toy car or truck. Take the toy apart. Draw and label diagrams of all the parts of your toy (4.7.1).

Extension

Challenge students to build cars and trucks, using a variety of materials such as Lego, Dacta, or other similar building sets.

Activity Centre

Provide a large variety of items such as old clocks, radios, toys, and so on, along with tools to take apart the items. Provide plenty of time for students to learn about these operating systems through manipulation and discovery.

Note: Safety is an issue if items have sharp parts. Check over selected objects carefully, and supervise this activity as required.

Assessment Suggestion

As students examine and take apart their toys, observe their ability to observe and record their observations on the activity sheet. Use the individual student observations sheet on page 16 to record results.

Date: ____________________ Name: ________________________________

Toy Systems

My toy is a ______________________________.

The parts of my toy are:

References for Teachers

Gear Up For Fun, Engineering Nature: The Art & Science of Natural & Built Worlds. Vancouver, BC: Association for the Promotion and Advancement of Science Education, 1997.

Science Works: An Ontario Science Centre Book of Experiments. Toronto: Government of Ontario, 1984.

Walshe, Bridget. *Engineering For Children: Structures, A Manual for Teachers*. Vancouver, BC: Association for the Promotion and Advancement of Science Education, 1991.

Zubrowski, Bernie. *Messing Around With Drinking Straw Construction*. Boston: Little, Brown & Company, 1981.

Unit 5
Daily and Seasonal Cycles

Books for Children

Barrett, Judi. *Animals Should Definitely NOT Wear Clothing.* New York: Aladdin, 1989.

Branley, Franklyn. *What Makes Day and Night.* New York: Crowell, 1986.

Brown, Margaret Wise. *Goodnight Moon.* New York: HarperCollins, 1991.

Chanko, Pamela. *Weather.* New York: Scholastic, 1998.

Gibbons, Gail. *The Seasons of Arnold's Apple Tree.* New York: Harcourt Brace Jovanovich, 1984.

Graves, Kimberlee. *See How It Grows.* Learn to Read Science Series, Level 1. Cypress, CA: Creative Teaching Press, 1994.

_____. *Pack a Picnic.* Learn to Read Science Series, Level 3. Cypress, CA: Creative Teaching Press, 1994.

Hurd, Edith T. *Day the Sun Danced.* New York: n.p., 1965.

McGuire, Richard. *Night Becomes Day.* New York: Viking, 1994.

Raffi. *One Light, One Sun.* New York: Crown Publishers, 1988.

Sendak, Maurice. *Chicken Soup With Rice; A Book of Months.* New York: Scholastic, 1962.

Silverstein, Shel. *The Giving Tree.* New York: HarperCollins Books for Children, 1987.

Tibo, Gilles. *Simon Welcomes Spring.* Montreal, PQ: Tundra Books, 1990.

_____. *Simon in Summer.* Montreal, PQ: Tundra Books, 1991.

_____. *Simon and the Wind.* Montreal, PQ: Tundra Books, 1991.

_____. *Simon and the Snowflakes.* Montreal, PQ: Tundra Books, 1988.

Yolen, Jane. *Owl Moon.* New York: Philomel Books, 1987.

Williams, Rozanne Lanczak. *How's the Weather?* Learn to Read Science Series, Level 1. Cypress, CA: Creative Teaching Press, 1994.

_____. *Round and Round the Seasons Go.* Learn to Read Science Series, Level 1. Cypress, CA: Creative Teaching Press, 1994.

_____. *The Four Seasons.* Learn to Read Science Series, Level 1. Cypress, CA: Creative Teaching Press, 1994.

_____. *What's the Weather Like Today?* Learn to Read Science Series, Level 2. Cypress, CA: Creative Teaching Press, 1994.

Creative Teaching Press books are available from Peguis Publishers, Winnipeg

Web Sites

- http://www.theideabox.com/

  An excellent site dedicated to the education of young children. Click on "Seasonal" to find activities, crafts, and projects for seasonal themes.

- http://www.geocities.com/Heartland/7134/Shadow/groundhog.htm

  An educational site for students and teachers. Learn about shadows, groundhogs, hibernation, and sundials. Includes suggested projects, puzzles, and an online quiz.

- http://www.owu.edu/~mggrote/pp/

  Project Primary is a collaborative effort between elementary teachers and professors from six different departments. Click on "Physics" to find information and activities on shadows, Earth, measuring shadows, and sundials.

- http://www.teelfamily.com/activites/snow/

  Activities for the study of snow. Includes snow crystals and how to examine them, snow activities, and snow links.

- http://www.planetdiary.com

  Planet Diary: an extensive web site for teachers and (older) students that records the events and phenomena that affect Earth and its residents. This site is updated weekly, with phenomena background information and activities. Includes Planet Diary calendar and archives.

- http://usatoday.com/weather/wworks0.htm

  How the Weather Works: graphics and text that explain weather phenomena, with extensive links.

- http://www.hao.ucar.edu/public/education/education.html

  NCAR High Altitude Observatory: "What is the sun?" "Can the sun be dangerous?" What is it made of?" "Why do we study the sun?" Answers to all these questions and more.

- http://www.nature-net/bears/

  The Bear Den: for teachers and students researching bears, this web site includes information on population and distribution, vital statistics, physical characteristics, diet and food sources, reproduction, mortality, and hibernation – for eight species of bears. The "Cub Den" is designed for young students, with "Ten Facts About Bears," and "Amazing Facts About Bears."

- http://www.kortright.org

  Located just ten minutes away from Metro Toronto, The Kortright Centre is Canada's largest environmental education centre – with over sixty different programs that have been designed to complement classroom curriculum.

Introduction

This unit focuses on changes as they occur daily, weekly, monthly, and throughout the seasons. Throughout the unit, students will be actively involved in investigations to do with changes over time. They will identify regular events, sequence events, and observe the characteristics of living things over periods of time.

The unit emphasizes several key concepts familiar to the grade-one program, such as sequencing events, identifying the days of the week, and identifying the characteristics of seasonal changes. For this unit, as with some others, you may find it more appropriate to focus on the suggested activities throughout the year, as opposed to teaching the unit in one block of time. This is especially true when studying the months and seasons.

Science Vocabulary

Throughout this unit, teachers should use, and encourage students to use, vocabulary such as: *days of the week, seasons of the year, yesterday, today, tomorrow, morning, afternoon, evening* and *night*.

Materials Required for the Unit

Classroom: chart paper, markers, scissors, mural paper, crayons, pastels, glue, coloured construction paper (various colours including black and yellow), paper clips, unlined index cards, masking tape, chalk, hand lenses, class journal (included), art paper, Plasticine, pencils

Books, Pictures, and Illustrations: picture books (for sequencing), picture cards for ordering events (included), magazines, pictures of things associated with daytime and nighttime (included), cards with days of the week on them, *The Seasons of Arnold's Apple Tree* (a book by Gail Gibbons), picture books and videos about animals during the seasons, *Weather* (a book by Pamela Chanko), clothing catalogues, books by Gilles Tibo (*Simon Welcomes Spring, Simon in Summer, Simon and the Wind, Simon and the Snowflakes*), small cards or cutouts (for weather, birthdays), weather cards (included)

Household: cloth, yarn, ribbon

Other: large monthly calendar, "One Light, One Sun" (a song by Raffi), shoeboxes, thermometers, aluminum foil, white cloth, red cellophane, cotton batting, outdoor thermometer, non-evergreen tree, laminated sign, plywood, sand, soil, rocks, grass, blue cellophane, cotton batting, materials for making bird feeders, bird feed, letter to parents/guardians (included), home survey (included), cassette tape of "A House Is a House" (by Fred Penner), paper bags, Hula-Hoops, clothing (e.g., mittens, scarves, jacket, shorts)

1 | Sequencing Events

Materials

- familiar picture books with obvious chronological sequences (have multiple copies available)
- picture cards (included) (5.1.1)
- scissors
- glue

Activity

Note: Prior to this activity, take two or more of the same picture book, separate the pages, and mount each page on Manila tag. You will use these mounted pictures for sequencing. Make sure one copy of the book remains intact.

Read one of the picture books aloud to the students, then discuss the events that occur in the story. Ask:

- What is the story about?
- Do you remember what happens first in the story?
- What happens next?
- How does the story end?

Display, in random order, one set of the mounted pictures for students to observe and discuss. Ask:

- What is happening in these pictures?
- Are the pictures in the same order that things happened in the story?
- Can you put them in the correct order so that they tell the story?

Have students order the pictures and describe the events aloud.

Repeat this activity with other picture books.

Activity Sheet

Directions to students:

Look carefully at the pictures. Cut out the pictures and glue them in the order they would have happened (5.1.1).

Extensions

- When reading stories or sequencing events in classroom activities, use graphic organizers or story maps to indicate the sequence. For example:

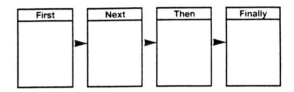

- Focus on living things and have students describe the sequence of events in the life cycle of, for example, a human (baby, toddler, child, teenager, adult) or a chicken (egg, chick, chicken).

- During cooking activities, sequence directions for making recipes.

Activity Centre

- Provide several picture books, along with randomly ordered pages from the books that you have mounted on Manila tag. Have students review the books and then arrange the mounted pages so they tell the same story that is in the book.

- Photograph students as they are doing a class activity, such as baking cookies or taking a nature walk. Have the students sequence the photographs to show the order in which the events occurred.

Ordering Events

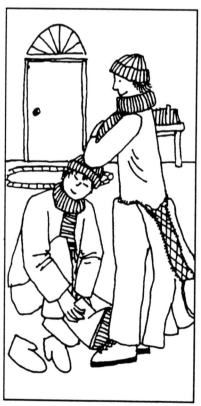

Ordering Events

Beginning	**Middle**	**End**

Beginning	**Middle**	**End**

2 | Daily Activities

Materials

- chart paper, markers
- scissors
- 3 large sheets of mural paper
- art supplies, such as crayons, pastels, magazines, glue, cloth, and coloured construction paper

Activity

As a class, discuss with students what they do before they come to school. Ask:

- What time do you wake up?
- What is the first thing you do in the morning?
- What do you do next?

Record students' activities on a chart titled Before School.

Discuss what happens during the school day. Ask:

- When does school start?
- What do you do first at school?
- What do you do next?

Record students' daily school activities on a chart titled At School.

Discuss activities done after school. Ask:

- What do you do when you leave school?
- What do you do before supper?
- What do you do after supper?
- What do you do before going to bed?

Record students' activities on a chart titled After School.

Divide the class into three groups and give each group one of the sheets of mural paper titled Before School, At School, or After School. Have the groups work on their murals

for ten minutes, creating pictures of activities that they do before school, at school, or after school. Rotate the groups twice. The result will be three murals to which all students have contributed. Ask:

- Can you see a pattern in the activities you do?
- Are there some activities you do every day?

Now discuss some of the things that students do on days when they are not at school. Ask:

- What do you do on a Saturday or Sunday?
- What do you do in the morning?
- What do you do in the afternoon?
- What do you do in the evening?
- What activities are the same on a school day and on a weekend day?
- What activities are different on a weekend day?

Use the ideas from this discussion to have students complete the activity sheet.

Activity Sheet

Directions to students:

Draw pictures and print words to show the activities that you do on a Saturday or Sunday (5.2.1).

Extensions

- Cut out the students' responses recorded from the three charts made during the above activity. Have students sequence the activities that they do before school, at school, and after school.

- Have students make a picture diary showing activities they have done over the course of a weekend. Students can present their diaries at school during sharing time.

▶

2

Activity Centre

Take photographs of students doing daily activities at school. Mount the photographs on construction paper and label each picture according to the activity being done. At the centre, have the students sequence these activities for a school day.

Assessment Suggestion

As students work together on the murals, observe their ability to work as a group. Record your observations on the cooperative skills teacher assessment sheet on page 21.

Date: _____

Name: _____

A Weekend Day

Morning	Afternoon	Evening

3 | Day and Night

Materials

- pictures of things associated with daytime and nighttime (included) (5.3.1-5.3.2) (Pictures can be enlarged for use with the class.)
- chart paper, markers
- 2 large pieces of construction paper (one black, one yellow) (Cut a half circle from the black construction paper and a half circle from the yellow construction paper. Tape the half circles together to create a sorting circle.)

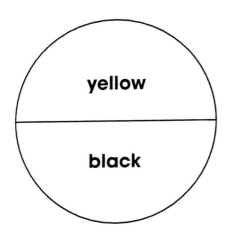

Activity

Have the students sit in a circle around the sorting circle. While the students watch, place a picture of an activity associated with daytime in the yellow section of the sorting circle. Then place a picture of an activity associated with nighttime in the black section of the sorting circle. Challenge the students to think about your sorting rule. In the meantime, add another picture to each of the half circles. When you have three pictures in each half of the sorting circle, ask:

- Why do you think these pictures are grouped together on the black paper?
- How are they the same?
- Why do you think these pictures are grouped together on the yellow paper?
- How are they the same?
- What is my sorting rule?

Give students an opportunity to share their ideas and to add pictures according to the daytime/nighttime sorting rule.

Once all pictures are sorted, lead a discussion about daytime/nighttime activities. Talk with the students about sights, sounds, and animals they might see in the daytime or the nighttime. Also discuss with students the jobs that humans do in the daytime and jobs that humans do in the nighttime.

Divide chart paper into two columns titled Daytime and Nighttime. Record the students' ideas on the chart.

Activity Sheet

Directions to students:

On the activity sheet, draw pictures and print words to show what you know about daytime and nighttime (5.3.3).

Extensions

- Read a variety of nonfiction books that focus on animal and plant behaviour both during the day and during the night (i.e., the behaviour of nocturnal animals and changes in certain plants and flowers).

- Give each student one whole and one half of a paper plate. On the whole paper plate, have them draw pictures of things that occur during the day (e.g., the sun shines, children are at school). On the half paper

►

plate, have them draw pictures that show things that happen at night (e.g., the moon, stars, and nocturnal animals come out).

Once their pictures are complete, have them attach the half plate over the whole plate with a fastener, making certain that the top half is moveable.

Students can now use their models to discuss daytime and nighttime activities.

■ Read the book *Goodnight Moon* by Margaret Wise Brown, and discuss the changes that occur as nighttime falls.

Things That Happen in the Daytime

Things That Happen at Nighttime

Day and Night

Daytime	Nighttime

4 | **Time Is Measured in Days and Weeks**

Materials

- 7 cards, each with a different day of the week printed on it
- 7 sheets of chart paper
- paper clips
- small, unlined index cards
- pencils, crayons
- masking tape
- glue
- scissors

Activity: Part One

Teach the song "The Days of the Week." This song is sung to the tune of "If You're Happy and You Know It."

> Sunday is the first day of the week.
> (clap, clap)
> Monday is the next day of the week.
> (clap, clap)
> Tuesday, Wednesday, Thursday, then Friday on to Saturday.
> Now you have the seven days of the week.
> (clap, clap)

While students sing the song, point to the name of each day on the cards and have students point to the correct card as they sing.

Have students make up circle stories using the days of the week. For example, the first student might say, "On Sunday, I went on a picnic and ate watermelon." The next student might add, "On Monday, I went on a picnic and ate egg salad sandwiches," and so on. It is not necessary to have the students repeat what previous students say, although this is another variation of the game that you may wish to try.

Other variations are:

- On Sunday I went to the store and bought _____.
- On Sunday I went to the park and played _____.
- On Sunday we got in the car and went _____.

Activity: Part Two

Note: This activity will take place over the course of a week. It is best to begin it on a Monday, so that Saturday's events and Sunday's events are still fresh in students' minds.

Start by reviewing the days of the week, using the cards and singing the song taught in part one of this activity.

Attach the Sunday card onto a piece of chart paper with a paper clip. Discuss activities that students did on Sunday, and record these on the chart paper. On the unlined index cards, have each student draw a picture of one activity they did on Sunday. Attach the drawings to the Sunday chart with masking tape.

Near the end of each school day, record activities students have done that day and have them draw pictures of those activities. Attach these drawings to chart paper headed by the day of the week.

Review the chart so students have a chance to share their routines with classmates. This activity also reinforces the concept of events that occur within a one-day period.

▶

4

Activity Sheet

Note: Enlarge the activity sheet to ledger size to provide students with more space for pictures.

Directions to students:

Cut out the names of the days of the week. Glue them in order, and draw a picture of your favourite activity for each day of the week (5.4.1).

Extension

Have students make zigzag picture books of something they do on each day of the week. To make a zigzag book, fold long strips of paper into eight sections in the following manner:

Use the first section as a title page. Students can then print the names of the days of the week at the top of each of the following sections. Encourage them to refer to the charts and cards for guidance. Now have students draw pictures of activities done on a particular day. Extend this activity for a week, adding pictures to the book each day. The book can be folded up like a fan, and unfolded to read.

Assessment Suggestion

Using the index cards with the names of the days of the week, have students sing the song, "The Days of the Week," and identify each day on the index cards. Then mix up the cards and have them sequence the days from Sunday to Saturday. Use the anecdotal record sheet on page 15 to record results.

Name: _____

Date: _____

Days of the Week

Tuesday	Saturday	Monday	Friday	Sunday	Wednesday	Thursday

5 | Months of the Year

Materials

- large monthly calendar (commercial or teacher-made)
- small cards or cutouts (to attach to the calendar for weather, birthdays, and other special events)
- glue, scissors

Activity

Note: This activity will run over a period of one month and should be continued throughout the year.

Introduce each month and discuss its name. Ask the students:

- What month is it today?
- What is the date when the month begins?

Stress the proper way to read the date; for example, "Today is Monday, October 4." Have students repeat the date on a daily basis. You may also begin to talk about yesterday and tomorrow in the same way. This should be a guided activity; do not expect students to do this independently.

Focus on the day's weather. Ask:

- What is the weather like today?
- It is sunny or cloudy?
- Is it snowing or raining?
- Is it windy or calm?

Add a weather card for the day.

Focus on upcoming events during the month. Ask:

- Does anyone have a birthday this month?
- Are there any special occasions this month?
- Do we have any field trips or special events this month?

Add birthday cards and special event cards.

During the month, continue to focus on upcoming events. Ask:

- How many days until [student's name] birthday?
- How many days are left in the month?

At the end of the month, review the weather, birthdays, and special events that occurred that month. (This reinforces the concept of events that occur in a one-month period.)

Use the monthly calendar to reinforce math concepts such as counting, skip counting, before and after, and patterns. On a regular basis, take opportunities to include these types of activities in your calendar study.

Activity Sheet

Directions to students

Record this month's activities and special events on the calendar (5.5.1).

Extensions

- The monthly calendars made by the students (5.5.1) can be part of a monthly newsletter that is sent home at the beginning of each month. Students can fill in daily events, special activities, field trips, and so on.

- Make a graph identifying students' birthdays each month.

- Read *Chicken Soup With Rice: A Book of Months* by Maurice Sendak to further focus on the months of the year.

- Over a period of one month, record the daily weather, then, as a class, construct a picture graph of the month's weather. Use the weather cards (5.5.2) for the picture graph.

Date: _____

Name: _____

Calendar for _____

Sunday	Monday	Tuesday	Wednesday	Thursday	Friday	Saturday

Weather Cards

sunny	sunny	sunny	sunny	sunny
cloudy	cloudy	cloudy	cloudy	cloudy
raining	raining	raining	raining	raining
snowing	snowing	snowing	snowing	snowing
windy	windy	windy	windy	windy
hot	hot	hot	hot	hot
cold	cold	cold	cold	cold

6 The Sun as a Source of Heat

Science Background Information for Teachers

The Sun is our closest star. It keeps Earth warm. The Sun's rays are called solar radiation. They are made of heat and light energy. These rays shine on Earth, travelling through the atmosphere and warming the air and Earth's land and oceans. Some of this radiation bounces or reflects off Earth, and is lost in space. But some is trapped in our atmosphere and it keeps us warm.

Materials

- lyrics, cassette, or CD of Raffi's song "One Light, One Sun"
- chart paper, markers
- 6 shoeboxes
- 6 thermometers
- various materials such as aluminum foil, white cloth, black construction paper, red cellophane, and cotton batting
- outdoor thermometer

Activity: Part One

Note: It is best to conduct this activity on a sunny day.

As an introduction to concepts about the sun, teach the students Raffi's song, "One Light, One Sun." Take the class outside to observe the Sun and its characteristics. (Instruct the students not to look directly at the Sun because it can damage the eyes.) Ask:

- Where is the Sun?
- Can you touch the Sun?
- What does the Sun look like from where you are?
- What do you think the Sun feels like?
- Is it hot or cold?
- How does the Sun help you?

Back in the classroom, title a chart What We Know About the Sun. Record students' ideas about how the Sun looks, and what the Sun does. Their ideas might include the following:

- The Sun is in the sky.
- The Sun looks round.
- The Sun is hot.
- The Sun gives us light.
- The Sun keeps us warm.
- You can get a sunburn if you are in the Sun too long.

Encourage the students to use terms related to the Sun, such as *rays, sunlight, sunshine, warm, hot, daytime, nighttime, sunrise,* and *sunset.*

Activity: Part Two

Note: You will introduce students to thermometers during this activity. It is not necessary for students to specifically read the temperature in degrees Celsius. However, students can begin to understand how thermometers work and how they show a rise and fall in temperature.

Focus on the Sun as a source of heat. Distribute the thermometers for students to observe and examine. (Encourage them not to touch the tip of the thermometer.) Ask:

- What are these called?
- What do you use them for?
- How do you use them?
- Why do you think I asked you not to touch the tip of the thermometer?

Have students examine the red liquid inside the thermometers and look at the calibrated numbers along the side of the liquid. Compare the various thermometers to determine if the red liquid is at the same point on each thermometer (the temperature on each should be close if students have not been touching the tips of the thermometers). Now have students hold the tips of the thermometers and observe closely to see what happens.

Take five of the shoeboxes and place one of the materials (such as aluminum foil or cotton batting) inside each. Leave the last shoebox empty. Display the shoeboxes. Have the students place a thermometer in each, then place the shoeboxes in direct sunlight. Ask:

- What do you think will happen to the thermometer in each shoebox?

Test the students' predictions by leaving the shoeboxes in the direct sunlight for several hours. Without removing the thermometers, have the students observe them. Ask:

- What has happened to the thermometers?
- Are all the thermometers showing the same temperature?
- Which thermometer is showing the highest temperature?
- Which thermometer is showing the lowest temperature?

Encourage the students to discuss the various materials and how each affected the thermometers.

As a follow-up to these activities, place an outdoor thermometer outside a classroom window so students can observe the changes in temperature during one day and from day to day.

Activity Sheet

Directions to students:

Record what you know about the Sun. Use pictures and words to show your ideas (5.6.1).

Extension

Have the students make suncatchers. You will need the following materials: newspaper, liquid starch, wax paper, white crayons, scissors, Styrofoam trays, paintbrushes, stapler, black construction paper, tape, and red, yellow, and orange tissue paper.

Have the students: cut out several 2.5 cm squares of tissue paper in each colour (or you can do this ahead of time). Spread out newspaper on a work surface and pour starch into the Styrofoam trays. Distribute a 22 cm x 30 cm piece of wax paper and two 22 cm x 30 cm pieces of black construction paper to each student. Set out a pile of the coloured tissue paper squares. Have each student "paint" liquid starch over the wax paper, then spread overlapping tissue squares on top of the wax paper, leaving a 2.5 cm-5 cm wide border around the wax paper. Next have the students paint over the squares with more starch.

When the tissue paper has dried, have the students use the white crayons to draw an outline of a sun on one of the sheets of black paper. Cut out the sun, then trace the outline onto the second sheet of black paper so that both are the same size. Cut the sun out of the second sheet. Leave the frame around both sheets uncut. Insert the decorated wax paper between the two black frames and staple together. Tape the suncatchers to the window.

What I Know About the Sun

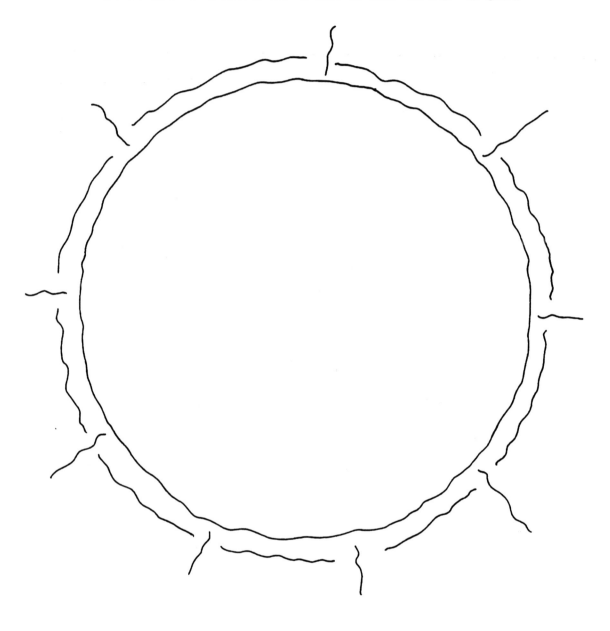

_____

_____

_____

_____

_____

7 | Shadows and the Position of the Sun

Science Background Information for Teachers

At the grade-one level, the focus of these activities should be on the observation of a shadow's change rather than on a scientific explanation as to why the change occurs. It is important to know that shadows are created by objects that block the Sun's rays. The changes in a shadow's length, width, and position occur because, as Earth rotates on its axis, the angle of the Sun's rays on an object changes.

Shadows can be measured and marked. Since Earth rotates counterclockwise, shadows are long, thin, and found west of the object early in the morning. As the Sun moves higher in the sky, shadows become shorter and wider. Shadows are shortest at noon, when the Sun is almost overhead. Later in the day, shadows lengthen again, and are found on the east side of the object.

Materials

Note: Materials will be provided for students working in groups.

- chalk
- yarn
- scissors
- chart paper
- tape
- markers

Activity

Note: Outdoor shadow activities must be done on sunny days. This activity should begin early in the school day.

Take students outside to the playground for a shadow walk. Ask them to observe the shadows cast by posts, basketball standards, flagpoles, buildings, and classmates. Then have them look at their own shadows. Ask:

- Does your shadow move when you move?
- Can you jump on your own shadow?
- What are the biggest and smallest shadows you can make with your body?
- Can you make your shadow touch a friend's shadow without touching your bodies together?

Now focus on shadows cast by still objects such as play structures and flagpoles. As you focus on each shadow, ask:

- What is the shape of the shadow?
- What causes a shadow?
- Where is the light coming from that makes the shadow?
- Do you think the shadow will change during the day?

Divide the class into working groups and provide each group with yarn, chalk, and scissors. Have each student mark an X on the pavement, print his or her name under the X, then stand on that spot.

Have the students trace the shape of one another's shadows with chalk. Now have them use the yarn to measure the length of the shadow, then cut the piece of yarn to that length. Ask:

- Do you think that the shadow will change if you come back to look at it later in the day?

Go back to the classroom and have students note the time of day. You do not have to use standard time measure. You can record times using intervals of the school day; for example, the beginning of the day, before recess, before lunch, after lunch, and before home time. Provide each group with a sheet of chart paper. Have them tape their piece of yarn onto the paper and record the time that the shadow was measured.

▶

Go outside several times during the day to observe the shadows and the position of the Sun. Each time, have the groups trace the new shadows with chalk and measure the length of the shadows with the yarn. Ask:

- How have the shadows changed?
- Are the shadows in different places?
- Are the shadows shorter or longer?
- Are the shadows bigger or smaller?
- Why do you think the shadows have changed?

After each observation, return to the classroom and have the groups tape their yarn onto the chart paper and record the time of day. At the end of the day, have the groups title their chart and present it to the class. During presentations, focus on their understanding of shadows. Ask:

- What creates shadows?
- Why do they change during the day?
- Do you think you would see shadows clearly on a cloudy day?

In their groups, have the students use their shoes to measure the length of each piece of yarn. Record these measurements on chart paper. Students can then transfer them onto their activity sheets.

Activity Sheet

Directions to students:

Record the time of day and the length of the shadow (number of shoes long) (5.7.1).

Extensions

- **Butterflies in a net:** You will need: several butterfly shapes cut from dark paper, a butterfly net (or stretch a pair of nylons over a bent coat hanger frame and tape

the end of the hanger). Tape the butterflies in a bright sunny window. The Sun should project the shapes onto the floor. Have the students capture (cover) the shadow butterflies with the net. Leave the net in place on top of the butterflies. Check on the butterflies later. As the Sun moves, the students will be amazed to see the butterflies escaping from the net. Ask the students to explain how this happened.

- Use a globe to introduce the idea of Earth's rotation. This will help the students to understand how Earth's position changes the position of shadows.

- Have students make paper plate sundials. Push a pencil through the centre of an upside-down paper plate. Take the sundial outside and push the pencil into the ground. The pencil will make a shadow on the plate. Students can mark the position of the shadow at this time, and several times during the day.

- Play shadow tag in the morning and afternoon. The object of the game is for students to try to touch one another's shadows. Encourage them to explain how the game changed when they played it at two different times of the day.

Date: _____ Name: _____

Shadows

Time of Day	Length of Shadow (Number of Shoes)

8 Seasonal Cycles

Note: "Tree watching" helps young students understand seasonal changes. Adopting a tree provides motivation for the study of trees. It is important for the students to look for changes above, around, and below the ground when investigating their tree.

Science Background Information for Teachers

Broadleaf, or deciduous, trees lose their leaves in the fall when the amount of sunlight lessens and the days become cooler. Winter brings frost, snow, and ice, and the ground becomes hard. The trees are bare and nearby plants are in a dormant stage. Tree growth slows down during this time. In spring, buds appear and we see new leaves. We also see new grass and flowers around the trees, and the soil and air temperature feel warmer under the trees. Trees remain green for most of the summer, but the soil and plant life beneath them will start to dry.

Note: The animals that habit or frequent the trees will change throughout the seasons as well.

Materials

- *The Seasons of Arnold's Apple Tree,* a book by Gail Gibbons (or any other book that focuses on living things throughout the seasons)
- 4 sheets of chart paper, markers
- non-evergreen tree for adoption
- ribbon or laminated sign (for marking tree)
- thermometers
- yarn, scissors
- paper bags
- hand lenses
- class journal for recording observations while visiting the adopted tree (a large scrapbook or sketchbook works well for this activity) (included) (5.8.1)
- paper and crayons for making tree rubbings
- pencils

Activity

Note: This activity is best done beginning in the fall, so that students can observe the changes throughout the seasons.

Begin by reading *The Seasons of Arnold's Apple Tree.* Discuss the events and ideas in the book. Ask:

- What do you know about seasons?
- What season is it now?
- What are the names of the four seasons?
- What is the order of the four seasons?
- What happens in spring?
- What happens in summer?
- What happens in fall?
- What happens in winter?
- Why do you think the seasons change?
- How does the weather change during the seasons?
- Why do you think the weather changes during the seasons?

Title the four sheets of chart paper with the names of the seasons. Have the students brainstorm words that can be associated with each season, and record these on the appropriate chart.

As part of their study on the seasons, explain to the students that, as a class, they will be adopting a tree to study during the school year. Ask:

- What happens to trees during the different seasons?
- Do all trees change?
- Which trees do not seem to change during the seasons?
- What do you think you could watch for as you study your tree?

Visit your adopted tree. Take pencils, ribbon, yarn, scissors, paper bags, thermometers, paper, crayons, and a class journal (optional:

▶

8

laminated sign). Designate the tree with ribbon or a laminated sign. Allow students time to investigate the tree and to draw a picture of it. Encourage the students to use their senses during their investigation of the tree. For example:

- Touch the bark, leaves, and ground around the tree.
- Observe and discuss the warmth of the tree and soil.
- Smell the bark, leaves, and soil.
- Listen to the sounds around the tree.
- Observe what is growing around the tree.

Have the students measure the circumference of the tree: wrap the yarn around the tree trunk, then cut the yarn. Place the yarn in the paper bag so it can be measured when you are back in the classroom.

Have the students measure the temperature of the air and soil around the tree and record these temperatures, along with the date, in the class journal.

Students can also collect a fallen leaf or twig, or make a crayon rubbing. These can be collected in the paper bags.

Back in the classroom, talk to the students about their visit to their adopted tree. Have them examine the twigs and leaves they have collected, or the rubbings they have made. Ask:

- What did the tree look like, feel like, and smell like?
- What sounds did you hear around the tree?

Tape the yarn that measured the circumference of the tree onto chart paper. Record the date that the tree was measured. Measure the circumference of the tree with yarn during each visit and add the results to the chart.

Begin a class graph of the air and soil temperatures around the tree.

Visit the tree every few weeks during the school year to observe changes. Each time, collect fallen twigs and sticks, and make rubbings; measure the tree's circumference; and take the temperature of the air and soil around the tree. As the seasons change, discuss the changes in the tree's characteristics and growth, and the temperatures of the soil and the air.

Activity Sheet A

Note: Use the black line master to create tree journals for the students. Students can use these to record their observations about the tree throughout the seasons.

Directions to students:

Use pictures and words to describe your adopted tree (5.8.1).

Activity Sheet B

Directions to students:

Draw pictures and use words to show how the weather and living things change from season to season (5.8.2).

Extensions

- Throughout the school year, read books, learn songs and chants, and discuss the seasonal changes that occur in trees, animals, and the weather.

- Play Four Corners, using words that have been brainstormed for each season. Place a card with the name of each season in each corner of the classroom or gym. Call out a word from one of the lists. Students then move to the corner of the room labelled with the season that the word relates to.

▶

8

Activity Centre

- Provide large sheets of mural paper, crayons, markers, pastels, as well as materials for collages such as pictures from magazines, cloth, twigs, leaves, and glue. During the school year, title one sheet of mural paper according to whatever season it is. Have students use the art supplies to create a mural depicting the weather and behaviour of living things during that season. As the season changes, make a mural to depict the new season.

- Make class big books depicting each season. Provide large sheets of paper, along with pencils and crayons. Have each student create one page for the book by writing and drawing depictions of the weather and the behaviour of living things during a particular season. Make a big book for each of the four seasons to place in the classroom or school library.

- On index cards, write the words that students brainstormed for each of the seasons (you can include pictures on these cards to support the students' reading). Have the students sort the words according to each season.

Assessment Suggestion

Using the rubric on page 19, identify five criteria related to the seasonal cycle. For example:

1. Identify and sequence the seasons.
2. Identify five words/characteristics associated with spring.
3. Identify five words/characteristics associated with summer.
4. Identify five words/characteristics associated with fall.
5. Identify five words/characteristics associated with winter.

During individual interviews, ask students to identify the seasons in order and to give words that remind them of each season. Record the results on the rubric.

Our Tree

Date: _____

Name: _____

Our Tree

Date: _____

Name: _____

Name: _____

Date: _____

The Seasons

Summer

Winter

Spring

Fall

9 | Animals During the Four Seasons

Materials

- variety of picture books, filmstrips, videos, and so on, about animals during the seasons
- 4 sheets of chart paper, markers
- art paper and pencils
- sheets of plywood (at least 60 cm square)
- Plasticine
- materials for making habitat models, such as sand, soil, twigs, rocks, and grass; blue cellophane (for water); cotton batting (for snow)
- various materials for making bird feeders (these will be determined by students)
- bird feed

Activity: Part One

Note: The first part of this activity challenges students to design a three-dimensional model depicting a specific animal's behaviour and habitat during the four seasons. The model is made on a sheet of plywood that is divided into four sections, one for each season. Students then construct a scene, much like a diorama, in each section of the model.

Read several books about animals throughout the seasons. Title the four sheets of chart paper with the headings Animals in Spring, Animals in Summer, Animals in Fall, and Animals in Winter. Record the students' ideas about animal behaviours during each of the seasons.

Divide the class into working groups and have each group select an animal to study. Have the groups use the activity sheet to plan a design for a model depicting their animal during the various seasons. For example, if a group is studying geese, they can show the geese flying north in the spring, living in a marsh in the summer, flying south in the fall, and living by a lake in the winter.

After the groups have designed their models, they can construct the models on the plywood board. First, have them divide the plywood board into four sections (to designate the four seasons). Next, have them make models of their animal with Plasticine. Then, let them use the materials to create their model.

Have the groups display their models to the rest of the class.

Activity Sheet A

Directions to students:

Design your model of an animal in spring, summer, fall, and winter and list materials you will need to construct your model (5.9.1).

Extension

Conduct specific research on black bears and polar bears. Focus on the animals' behaviours throughout the seasons, as well as on what these animals do to protect themselves as the seasons change.

Activity: Part Two

Note: This activity involves students using the design process to construct a useful item. The design process is as follows:
1. Identify a need.
2. Create a plan.
3. Develop a product.
4. Communicate the result.
The design process also involves both research and experimentation.

For this activity, challenge students to design and construct a bird feeder. Start by focusing on birds that stay in your area during the winter. Ask:

- What types of birds have you seen here in the winter?
- What do they look like?
- What do birds eat in the winter?

9

- How do you think they get their food?
- Do you think there is enough food for them in the winter?
- Why not?
- How can you help birds stay healthy in the winter months?

Challenge the students to make bird feeders to help birds survive the winter. Ask:

- What materials could you use to make a bird feeder?
- Where would you put a bird feeder?
- What else would you need? (bird seed)

Give the students plenty of time to plan and design a bird feeder. Have them use the activity sheet to draw their design and list the materials they will need. Encourage them to use their own ideas, but provide support at the same time. Bird feeders can be made from milk cartons, large plastic pop bottles, pieces of wood, or recycled plastic containers. Remind students that they will need to plan a way of attaching the bird feeder to a tree, fence, or other structure. They should also ensure that the food is sheltered with a roof of some sort so it does not get buried in snow.

Once all students have constructed their bird feeders, have them present them to the class and then take the feeder home where they can observe birds feeding.

Activity Sheet B

Directions to students:

Plan and design your bird feeder. Draw a picture of the bird feeder and list all of the materials you will need to make it (5.9.2).

Extension

Encourage students to think of other ways that they can help animals survive during the seasons. Some suggestions are: filling bird baths with water, making a doghouse, keeping a dish outdoors filled with water for dogs and cats.

Assessment Suggestion

Assess the students' ability to use the design process when making the bird feeders. As they work, observe their ability to make a plan (refer to their activity sheets), construct the bird feeders, and present them to the class. Use the individual student observations sheet on page 16 to record results.

Date: _____   **Name:** _____

_____

_____

_____

Our Animal _____

Spring	Summer
Fall	Winter

Materials we will need:

_____   _____

_____   _____

Date: _____   Name: _____

My Bird Feeder

My Plan

Materials I will need:

_____   _____

_____   _____

_____   _____

_____   _____

_____   _____

10 Clothing Throughout the Seasons

Materials

- *Simon Welcomes Spring, Simon in Summer, Simon and the Wind*, and *Simon and the Snowflakes*, books by Gilles Tibo (or other books that depict weather and clothing throughout the seasons)
- an assortment of clothing (e.g., shorts, hooded jacket, mittens, scarf)
- large index cards labelled with the names of the seasons
- 4 Hula-Hoops
- magazines and department store catalogues with adult and children's clothing
- scissors, glue

Activity

Read and discuss the four books about how Simon welcomes the seasons. For each story, ask:

- What season is it now?
- How do you know what season it is?
- What was Simon doing?
- Were there any animals?
- Were there any plants?
- What clothes was Simon wearing?

Have the students examine the assortment of clothing, using the ideas from the books. Encourage them to discuss during which season each item of clothing is worn.

Have the students sit in a circle. Place the four Hula-Hoops in the middle of the group. Set the four index cards labelled with the names of the seasons next to the Hula-Hoops. Encourage students to sort the clothing according to the season in which the clothes are worn.

Note: Some clothing items may be worn during more than one season, so this is an opportunity to introduce intersecting sets, much like on an intersecting Venn diagram.

Discuss health and safety concerns related to dressing appropriately for the weather. Ask:

- What might happen if you do not wear mittens on a very cold day?
- How would you feel if you wore a big parka on a hot summer day?
- What should you wear when it is raining?

Throughout the school year, encourage students to dress in a manner that is appropriate for the weather and the season.

Activity Sheet

Note: Make two copies of the activity sheet for each student.

Directions to students:

At the top of each column print the name of one of the four seasons. Cut out pictures of clothing from catalogues and magazines. Glue them onto the pages to show the season in which you would wear the clothing (5.10.1).

Extensions

- Have a Seasons' Fashion Show. Ask each student to bring an outfit appropriate for one of the four seasons. Provide musical accompaniment and, as students display their outfits, have the class guess in which season the clothing would be worn. As a modification of this activity, have a Clothing Parade at the beginning of each season. Have students bring an appropriate outfit for the season that is just beginning, and parade throughout the school to remind other students of how to dress appropriately for the weather.

- Read the book *Animals Should Definitely NOT Wear Clothing* by Judi Barrett.

▶

10

Activity Centre

Dramatic Play Centre: Provide an assortment of clothing for the various seasons. Encourage students to dress for a season and dramatize activities done during that particular season. For example, dress for winter and pretend to skate or build a snowman. Provide different colours of clothing so that students learn that dark colours should not be worn in warmer weather because dark colours absorb heat. This can be related back to where the thermometers were placed in shoeboxes lined with various materials (see page 208).

Clothing Throughout the Seasons

Season	Season
_____	_____

11 | Activities Throughout the Seasons

Materials

- *Weather*, a book by Pamela Chanko (or any other book that focuses on seasonal changes)
- 4 sheets of chart paper, markers

Activity

Begin by reading the book *Weather,* and discussing how weather changes throughout the seasons. Ask:

- What is the weather like in summer/spring/fall/winter?
- What do humans do when it is windy/rainy/snowy/hot/cold?
- What do they wear?
- What activities do humans do in different seasons?
- How do humans get ready for the next season? (Some examples are: they get out seasonal clothes, prepare their yards, put on/take off storm windows, tune up snowblowers and lawnmowers.)

Title the four sheets of chart paper with the headings Humans in Spring, Humans in Summer, Humans in Fall, and Humans in Winter. Discuss, then record, the activities in which students and their families participate during the four seasons. Discuss students' favourite seasonal activities, and be sure to include these ideas on the charts.

Focus on how humans can do certain activities out of season. Ask:

- What would you do if you wanted to go swimming in the middle of winter?
- Would you go to the beach?
- Where else could you go?
- What if you wanted to ice skate during the summer?

- Would there be ice outside?
- Where else could you go?

To reinforce the various activities done during the seasons, play a version of charades. Have a student select an activity from one of the charts (with your assistance if required) and act out that activity for the class (e.g., downhill skiing, baseball, shovelling snow, raking leaves). Challenge the rest of the class to guess the activity and identify the season.

Activity Sheet

Directions to students:

Draw a picture and use words to describe your favourite activities in spring, summer, fall, and winter (5.11.1).

Extensions

- Discuss various weather extremes such as snowstorms, tornadoes, rain/floods, heat waves, lightning storms, and so on. Have students review how to stay safe in these conditions and discuss those who help us prepare for these weather extremes (examples are snowplough operators, meteorologists, news reporters).

- Make a class graph of the students' favourite activities during each of the seasons.

- Have students complete cloze sentences or rebus sentences that tell what they like best about each season. For example:

  I like _____ because _____.

  (I like winter because I play hockey.)

  I ♡ summer because I like to play ⬭.

226

Hands-On Science • Level 1

Date: _____

Name: _____

My Favourite Activities

Summer

Winter

Spring

Fall

12 | Shelter Throughout the Seasons

Materials

- lyrics or cassette tape of Fred Penner's song, "A House Is a House"
- parent letter (included) (5.12.1)
- home survey (included) (5.12.2)

Activity

Begin by teaching the students Fred Penner's song, "A House Is a House." Discuss the importance and features of houses. Ask:

- What features of a house keep you warm?
- What features of a house keep you safe?
- What features of a house allow you to see?
- What features of a house allow you to see, day and night?
- What features of a house protect you from the wind, rain, or snow?
- What features of a house allow you to keep and cook food to eat all year?

Explain to the students that they are going to conduct a study of where they live to find out how their homes shelter and protect them. With the students, review the letter to parents/guardian and the home survey so they understand the activity.

Send the letter and survey home with the students, and allow them ample time to work on this project. Encourage them to have their parents/guardians and siblings involved in the activity.

Once all students have completed the project, have them present their diagrams and information to the class for discussion and questions.

Activity Sheet

Directions to students:

Take the survey home. Use it to record the ways in which your home keeps you and your family healthy and safe (5.12.2).

Activity Centre

Provide students with a variety of building materials such as blocks, boxes, and Lego. Have them build houses, apartment blocks, bus shelters, picnic/park shelters, and other structures that protect people. Encourage them to include features of structures such as doors, windows, and so on.

Dear Parents/Guardians:

In science, we have been talking about the amount of light and precipitation in each season, as well as seasonal changes in temperature.

We are now studying about the features of homes that help us to adapt to changing weather conditions throughout the seasons.

Please help your child investigate your home and identify those features that keep you sheltered and comfortable (for example, furnace, lights, air conditioner, fans, windows, and curtains).

Thank you for your cooperation in this project.

Yours truly,

Dear Parents/Guardians:

In science, we have been talking about the amount of light and precipitation in each season, as well as seasonal changes in temperature.

We are now studying about the features of homes that help us to adapt to changing weather conditions throughout the seasons.

Please help your child investigate your home and identify those features that keep you sheltered and comfortable (for example, furnace, lights, air conditioner, fans, windows, and curtains).

Thank you for your cooperation in this project.

Yours truly,

Date: _____   Name: _____

Home Survey

Things in my home that help keep us warm in winter:

```
+------------------------+------------------------+
|                        |  _____ |
|                        |                        |
|                        |  _____ |
|                        |                        |
|                        |  _____ |
|                        |                        |
|                        |  _____ |
|                        |                        |
|                        |  _____ |
|                        |                        |
|                        |  _____ |
+------------------------+------------------------+
```

Things in my home that help keep us cool in summer:

```
+------------------------+------------------------+
|                        |  _____ |
|                        |                        |
|                        |  _____ |
|                        |                        |
|                        |  _____ |
|                        |                        |
|                        |  _____ |
|                        |                        |
|                        |  _____ |
|                        |                        |
|                        |  _____ |
+------------------------+------------------------+
```

Date: _____ Name: _____

Home Survey

Things in my home that protect us from snow, rain, and wind:

	_____
	_____
	_____
	_____
	_____
	_____

Other things that I learned about my home:

	_____
	_____
	_____
	_____
	_____

References for Teachers

Bosak, Susan. *Science Is...* . Richmond Hill, ON: Scholastic, 1991.

Butzow, Carol, and John Butzow. *Science Through Children's Literature*. Englewood: Teacher Ideas Press, 1989.

Liddelow, Lorelei. *Talk With Me*. Winnipeg, MB: Peguis Publishers, 1990.

McCracken, Robert, and Marlene McCracken. *Winter*. Themes. Winnipeg, MB: Peguis Publishers, 1987.

_____. *Fall*. Themes. Winnipeg, MB: Peguis Publishers, 1987.

_____. *Spring*. Themes. Winnipeg, MB: Peguis Publishers, 1987

Tompert, Ann. *Nothing Sticks Like a Shadow*. Boston: Houghton Mifflin, 1984.